Wide Bandgap Semiconductor Based Micro/Nano Devices

Wide Bandgap Semiconductor Based Micro/Nano Devices

Special Issue Editor

Jung-Hun Seo

MDPI • Basel • Beijing • Wuhan • Barcelona • Belgrade

MDPI

Special Issue Editor
Jung-Hun Seo
University at Buffalo, the State University of New York
USA

Editorial Office
MDPI
St. Alban-Anlage 66
4052 Basel, Switzerland

This is a reprint of articles from the Special Issue published online in the open access journal *Micromachines* (ISSN 2072-666X) from 2018 to 2019 (available at: https://www.mdpi.com/journal/micromachines/special_issues/Semiconductor_Materials_Micro_nano_Devices)

For citation purposes, cite each article independently as indicated on the article page online and as indicated below:

LastName, A.A.; LastName, B.B.; LastName, C.C. Article Title. *Journal Name* **Year**, *Article Number*, Page Range.

ISBN 978-3-03897-842-8 (Pbk)
ISBN 978-3-03897-843-5 (PDF)

Contents

About the Special Issue Editor

Jung-Hun Seo received his BS degree in electronics and electrical engineering from Korea University, Seoul, Republic of Korea, in 2006. He received his MS and PhD degrees in Electrical and Computer Engineering from University of Wisconsin-Madison in 2011 and 2015, respectively. Since 2016, he has been an assistant professor at the Department of Materials Design and Innovation, University at Buffalo, the state university of New York. He is the author or coauthor of more than 80 peer-reviewed papers, book chapters, and patents. His research interests mainly focus on the synthesis of low dimensional wide bandgap semiconductors toward high performance flexible electronics and optoelectronics. Also, he is working on various high frequency and high power devices based on wide bandgap semiconductors.

micromachines

MDPI

Editorial

Editorial for the Special Issue on Wide Bandgap Semiconductor Based Micro/Nano Devices

Jung-Hun Seo

Department of Materials Design and Innovation, University at Buffalo, The State University of New York, Buffalo, NY 14260, USA; junghuns@buffalo.edu

Received: 20 March 2019; Accepted: 25 March 2019; Published: 26 March 2019

check for updates

While conventional group IV or III-V based device technologies have reached their technical limitations (e.g., limited detection wavelength range or low power handling capability), wide bandgap (WBG) semiconductors which have band-gaps greater than 3 eV have gained significant attention in recent years as a key semiconductor material in high-performance optoelectronic and electronic devices [1,2]. These WBG semiconductors have various definitive advantages for optoelectronic and electronic applications due to their large bandgap energy. WBG energy is suitable to absorb or emit ultraviolet (UV) light in optoelectronic devices [3]. It also provides a higher electric breakdown field, which allows electronic devices to possess higher breakdown voltages [4].

In this Special Issue, 13 papers published, including various AlGaN/GaN, SiC, and WO_3 based devices. More than half of papers reported recent progress on AlGaN/GaN high electron mobility transistors (HEMTs) and light emitting diodes (LEDs). Wojtasiak et al., and Sun et al, reported a structural modification of AlGaN/GaN HEMTs to improve turn-on voltage, contact resistance, and on-resistance [5]. Huang et al. investigated high-temperature characteristics of AlGaN/GaN HEMTs and successfully established the thermal model [6]. Mao et al. and Li et al. simulated AlGaN/GaN HEMTs with a large signal model to investigate the kink-effect [7,8]. All of these efforts toward AlGaN/GaN HEMTs enable readers to understand current issues in AlGaN/GaN HEMTs and offer various experimental and theoretical solutions. Beside transistor works, flip-chip GaN LEDs that were combined with TiO_2/SiO_2 distributed Bragg reflectors (DBRs) was reported by Zhou et al [9]. An improved GaN HEMTs and their microwave performance by employing the asymmetric power-combining was reported by Kim et al [10]. Along with another GaN LED built on a modified micron-size patterned sapphire substrate by Hsu et al. [11]. These GaN LED works are also guided broad readers in the field of optoelectronics and biomedical areas toward future high-performance optogenetics and photonics applications. Also, Sun et al. reported an enhanced AlGaN/GaN Schottky Barrier by engineering the structure of the diode [12].

In addition to $Al_xGa_{1-x}N$ system, two SiC simulation efforts have been made by Huang et al. and Jia et al. Huang. They focused on the improvement of higher added efficiency (PAE) factor in 4H-SiC metal semiconductor field effect transistors and breakdown voltage of 4H-SiC diodes, respectively [13,14].

Besides popular $Al_xGa_{1-x}N$ and SiC-based applications, three papers report InGaZnO thin-film transistors (TFTs), Si/GaP one-transistor dynamic random-access memory (1T DRAM), and WO_3 thin-film. Zhou et al. investigated a stress tolerance of InGaZnO TFTs with a SiO_2 or Al_2O_3 passivation layer which shows a stable positive bias during the operation [15]. Kim et al. simulated a novel 1T DRAM design by inserting a GaP pillar which showed a stable high-temperature operation [16]. Finally, Zhang et al. reported the changes of the optical bandgap in Tungsten trioxide by thermal annealing which can be used for various electrochromic devices [17].

To the end, I would like to take this opportunity to thank all the authors for submitting their papers to this special issue. I also want to thank all the reviewers for dedicating their time and helping to improve the quality of the submitted papers.

References

1. Kim, M.; Seo, J.-H.; Singisetti, U.; Ma, Z. Recent advances in free-standing single crystalline wide band-gap semiconductors and their applications: GaN, SiC, ZnO, β-Ga$_2$O$_3$, and diamond. *J. Mater. Chem. C* **2017**, *5*, 8338–8354. [CrossRef]

2. Swinnich, E.; Dave, Y.J.; Pitman, E.B.; Broderick, S.; Mazumder, B.; Seo, J.-H. Prediction of optical band gap of β-(Al$_x$Ga$_{1-x}$)$_2$O$_3$ using material informatics. *Mater. Discov.* **2018**, *11*, 1–5. [CrossRef]

3. Liu, D.; Cho, S.J.; Park, J.; Gong, J.; Seo, J.-H.; Dalmau, R.; Zhao, D.; Kim, K.; Kim, M.; Kalapala, A.R.K.; et al. 226 nm AlGaN/AlN UV LEDs using p-type Si for hole injection and UV reflection. *Appl. Phys. Lett.* **2018**, *113*, 011111. [CrossRef]

4. Swinnich, E.; Hasan, M.N.; Zeng, K.; Dove, Y.; Singisetti, U.; Mazumder, B.; Seo, J.-H. Flexible β-Ga$_2$O$_3$ Nanomembrane Schottky Barrier Diodes. *Adv. Electron. Mater.* **2019**, *5*, 1800714. [CrossRef]

5. Wojtasiak, W.; Góralczyk, M.; Gryglewski, D.; Zając, M.; Kucharski, R.; Prystawko, P.; Piotrowska, A.; Ekielski, M.; Kamińska, E.; Taube, A.; et al. AlGaN/GaN High Electron Mobility Transistors on Semi-Insulating Ammono-GaN Substrates with Regrown Ohmic Contacts. *Micromachines* **2018**, *9*, 546. [CrossRef] [PubMed]

6. Huang, H.; Li, F.; Sun, Z.; Cao, Y. Model Development for Threshold Voltage Stability Dependent on High Temperature Operations in Wide-Bandgap GaN-Based HEMT Power Devices. *Micromachines* **2018**, *9*, 658. [CrossRef] [PubMed]

7. Mao, S.; Xu, Y. Investigation on the I–V Kink Effect in Large Signal Modeling of AlGaN/GaN HEMTs. *Micromachines* **2018**, *9*, 571. [CrossRef] [PubMed]

8. Li, J.; Mao, S.; Xu, Y.; Zhao, X.; Wang, W.; Guo, F.; Zhang, Q.; Wu, Y.; Zhang, B.; Chen, T.; et al. An Improved Large Signal Model for 0.1 μm AlGaN/GaN High Electron Mobility Transistors (HEMTs) Process and Its Applications in Practical Monolithic Microwave Integrated Circuit (MMIC) Design in W band. *Micromachines* **2018**, *9*, 396. [CrossRef] [PubMed]

9. Zhou, S.; Xu, H.; Liu, M.; Liu, X.; Zhao, J.; Li, N.; Liu, S. Effect of Dielectric Distributed Bragg Reflector on Electrical and Optical Properties of GaN-Based Flip-Chip Light-Emitting Diodes. *Micromachines* **2018**, *9*, 650. [CrossRef] [PubMed]

10. Kim, S.; Lee, M.-P.; Hong, S.-J.; Kim, D.-W. Ku-Band 50 W GaN HEMT Power Amplifier Using Asymmetric Power Combining of Transistor Cells. *Micromachines* **2018**, *9*, 619. [CrossRef] [PubMed]

11. Hsu, W.-Y.; Lian, Y.-C.; Wu, P.-Y.; Yong, W.-M.; Sheu, J.-K.; Lin, K.-L.; Wu, Y.S. Suppressing the initial growth of sidewall GaN by modifying micron-sized patterned sapphire substrate with H$_3$PO$_4$-based etchant. *Micromachines* **2018**, *9*, 622. [CrossRef] [PubMed]

12. Sun, Y.; Wang, Y.; Tang, J.; Wang, W.; Huang, Y.; Kuang, X. A Breakdown Enhanced AlGaN/GaN Schottky Barrier Diode with the T-Anode Position Deep into the Bottom Buffer Layer. *Micromachines* **2019**, *10*, 91. [CrossRef] [PubMed]

13. Huang, Y.; Wang, Y.; Kuang, X.; Wang, W.; Tang, J.; Sun, Y. Step-Double-Zone-JTE for SiC Devices with Increased Tolerance to JTE Dose and Surface Charges. *Micromachines* **2018**, *9*, 610. [CrossRef] [PubMed]

14. Jia, H.; Hu, M.; Zhu, S. An Improved UU-MESFET with High Power Added Efficiency. *Micromachines* **2018**, *9*, 573. [CrossRef] [PubMed]

15. Zhou, Y.; Dong, C. Influence of Passivation Layers on Positive Gate Bias-Stress Stability of Amorphous InGaZnO Thin-Film Transistors. *Micromachines* **2018**, *9*, 603. [CrossRef] [PubMed]

Micromachines **2019**, *10*, 213

16. Kim, M.; Ha, J.; Kwon, I.; Han, J.-H.; Cho, S.; Cho, I. A Novel One-Transistor Dynamic Random-Access Memory (1T DRAM) Featuring Partially Inserted Wide-Bandgap Double Barriers for High-Temperature Applications. *Micromachines* **2018**, *9*, 581. [CrossRef] [PubMed]

17. Zhang, G.; Lu, K.; Zhang, X.; Yuan, W.; Shi, M.; Ning, H.; Tao, R.; Liu, X.; Yao, R.; Peng, J. Effects of Annealing Temperature on Optical Band Gap of Sol-gel Tungsten Trioxide Films. *Micromachines* **2018**, *9*, 377. [CrossRef] [PubMed]

micromachines

Article

Effects of Annealing Temperature on Optical Band Gap of Sol-gel Tungsten Trioxide Films

Guanguang Zhang, Kuankuan Lu, Xiaochen Zhang, Weijian Yuan, Muyang Shi, Honglong Ning *, Ruiqiang Tao, Xianzhe Liu, Rihui Yao * and Junbiao Peng

Institute of Polymer Optoelectronic Materials and Devices, State Key Laboratory of Luminescent Materials and Devices, South China University of Technology, Guangzhou 510640, China;
msgg-zhang@mail.scut.edu.cn (G.Z.); mskk-lu@mail.scut.edu.cn (K.L.); mszhang_xc@mail.scut.edu.cn (X.Z.); 201430320366@mail.scut.edu.cn (W.Y.); 201430320229@mail.scut.edu.cn (M.S.); 201510102158@mail.scut.edu.cn (R.T.); msliuxianzhe@mail.scut.edu.cn (X.L.); psjbpeng@scut.edu.cn (J.P.)
* Correspondence: ninghl@scut.edu.cn (H.N.); yaorihui@scut.edu.cn (R.Y.); Tel.: +86-20-8711-4525 (H.N.)

Received: 6 July 2018; Accepted: 25 July 2018; Published: 30 July 2018

Abstract: Tungsten trioxide (WO_3) is a wide band gap semiconductor material that is used as an important electrochromic layer in electrochromic devices. In this work, the effects of the annealing temperature on the optical band gap of sol-gel WO_3 films were investigated. X-ray Diffraction (XRD) showed that WO_3 films were amorphous after being annealed at 100 °C, 200 °C and 300 °C, respectively, but became crystallized at 400 °C and 500 °C. An atomic force microscope (AFM) showed that the crystalline WO_3 films were rougher than the amorphous WO_3 films (annealed at 200 °C and 300 °C). An ultraviolet spectrophotometer showed that the optical band gap of the WO_3 films decreased from 3.62 eV to 3.30 eV with the increase in the annealing temperature. When the Li^+ was injected into WO_3 film in the electrochromic reaction, the optical band gap of the WO_3 films decreased. The correlation between the optical band gap and the electrical properties of the WO_3 films was found in the electrochromic test by analyzing the change in the response time and the current density. The decrease in the optical band gap demonstrates that the conductivity increases with the corresponding increase in the annealing temperature.

Keywords: optical band gap; tungsten trioxide film; annealing temperature; electrochromism

1. Introduction

Tungsten trioxide (WO_3) is an important indirect band gap semiconductor material [1]. It is used as a functional layer in the applications of gas sensors [2], photocatalysis [3], solar cells [4], water splitting [5] and electrochromism [6]. Electrochromic devices, such as smart windows [7], can meet the market demand of energy-saving devices. Since WO_3's electrochromic properties were found, researchers have widely studied WO_3-based electrochromic thin films and device applications [8].

There are various choices for preparing WO_3 films with the development of thin film technology. These include sputtering [9], chemical vapor deposition [10], vacuum evaporation [11], and sol-gel [12], among others. Currently, magnetron sputtering is a commercial technology that is used to prepare WO_3 films due to its uniformity of film and reliability. However, the high cost issue and problems in preparing large-size devices cannot be ignored. The sol-gel method is a feasible technology for reducing the cost even, though there are still some problems at the present stage, such as film inhomogeneity and poor process repeatability, among others. With the development of new sol-gel techniques, such as inkjet printing [13], sol-gel technology is promising for commercial applications in the future.

The optical and electrical properties of WO_3 film are related to the parameters of the sol-gel technique, such as the solvent [14], precursor [15] and annealing temperature [16], among others.

In previous work, there was a significant difference in the band gap of the WO$_3$ films obtained using different processes [17,18]. Therefore, it is worthwhile to launch further investigations into the relationship between band gap and the optical and electrical properties of WO$_3$ films, especially in regards to electrochromic properties. In this paper, we conducted a study on the optical band gap of WO$_3$ films with different annealing temperatures. The crystallinity, response time morphology and conductivity were also analyzed together. A correlation between the optical band gap and the electrical properties (conductivity) was found.

2. Materials and Methods

Tungsten powder (W, 99.5% metals basis, Macklin Biochemical Co. Ltd, Shanghai, China) and hydrogen peroxide (H$_2$O$_2$, Hydrogen peroxide 30%, Guangzhou chemical regent factory, Guangzhou, China) were mixed in a beaker with a water bath at 25 °C. After the reaction finished, an evaporative concentration treatment (at 150 °C) was conducted to remove the surplus H$_2$O$_2$. Finally, an appropriate anhydrous ethanol was added into the concentrated solution and the mixed solution was sealed and stirred for 3 h at 70 °C to obtain the sol-gel. A spin coating technique was used to prepare the WO$_3$ films (around 80 nm) on the indium tin oxide (ITO) glass. The thickness of the WO$_3$ film was optimized and controlled by the concentration of solution and spin coating parameters and it had an important influence on the electrochromic transmittance modulation ability [19]. In this work, the annealing temperature was focused on and other unrelated variables (sol concentration, spin coating parameters, substrate, electrolyte, etc.) were controlled. These as-deposited films were annealed at 100 °C, 200 °C, 300 °C, 400 °C and 500 °C for 60 min, respectively.

The crystallization of the film was analyzed by X-ray Diffraction (XRD, PANalytical Empyrean DY1577, PANalytical, Almelo, The Netherlands). The surface morphology was measured by atomic force microscopy (AFM, Being Nano-Instruments BY3000 Being Nano-Instruments, Beijing, China). The electrochromic test was performed using 0.8 mol/L of lithium perchlorate/propylene carbonate (LiClO$_4$/PC) electrolyte and an electrode gap (~1 mm). The transmission spectra were measured by an Ultraviolet spectrophotometer (SHIMADZU UV2600, SHIMADZU, Tokyo, Japan), with ITO glass (Optical band gap: >4 eV) acting as a blank. The current of the electrochromic test was recorded by an electrochemical workstation (CH Instruments CHI600E, CH Instruments, Shanghai, China). The relationship between the change of transmittance and the time was measured by a micro-spectrometer (Morpho PG2000, Morpho, Shanghai, China), with ITO glass acting as a blank.

3. Results and Discussions

Figure 1 illustrates the X-ray patterns of the WO$_3$ films that were annealed at different temperatures. The crystalline structures of these films were further analyzed using Jade 6.0 and PDF#30-1387 and PDF#41-0905. In Figure 1a, there are diffraction peaks of WO$_3$ at the patterns of the WO$_3$ films annealed at 400 °C and 500 °C, which demonstrate that these films transformed from amorphous to crystalline when the annealing temperature is higher than 400 °C. Furthermore, the change of crystalline structure was analyzed in Figure 1b. The characteristic diffraction peaks of WO$_3$ films (400 °C) indicate that the WO$_3$ films initially transform from an amorphous to a monoclinic structure. When the annealing temperature reached 500 °C, there was only one diffraction peak of the WO$_3$ film in the range of 2θ (22° to 26°), which demonstrated that the monoclinic structure of the WO$_3$ films turned into a cubic structure. Strictly speaking, the stoichiometric ratio of tungsten and oxygen was not fully satisfied with 1:3. Therefore, there was an oxygen vacancy which influenced the optical and electrical properties of WO$_3$ films [20].

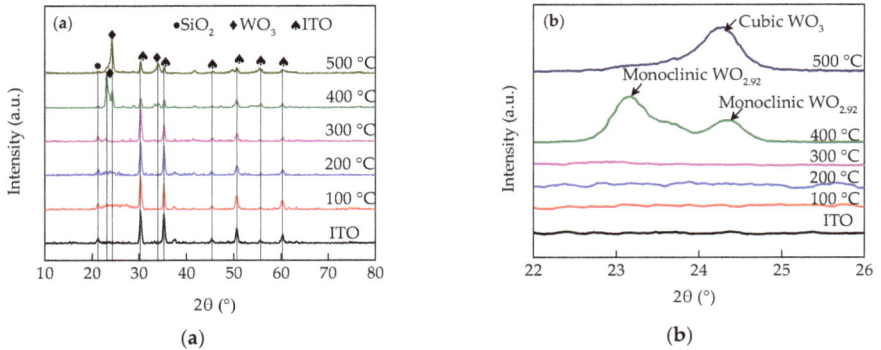

Figure 1. X-ray patterns of WO₃ films annealed at different temperature. (**a**) The XRD patterns in a large range of 2θ (10° to 80°); (**b**) The XRD patterns in a small range of 2θ (22° to 26°). The amorphous WO₃ transformed into monoclinic structure and cubic structure at 400 °C and 500 °C, respectively.

The surface morphology of these films was measured by AFM and the results are shown in Figure 2. Figure 2f shows a comparison of the roughness of these films at different annealing temperatures. The surface of the WO₃ film that was annealed at 100 °C is rougher than other films, which is confirmed by Figure 2a and its roughness. In this work, the solvent of sol was ethanol and water, which has a boiling point of around 80 °C. The 100 °C annealing treatment can remove the solvent, but it is not enough to remove the bound water in the tungsten acid [21]. In addition, solvent evaporation can cause defects in the surface, such as pores [22], and there is not enough energy to reduce these defects during annealing treatment. Therefore, among these samples, the WO₃ film annealed at 100 °C had the highest roughness.

The roughness of the films annealed at 200 °C and 300 °C was around 1.9 nm, which is less than that of the films (around 3.3 nm) annealed at 400 °C and 500 °C. This demonstrated that the crystalline film was rougher than the amorphous film because of its grain growth at a high temperature. The change in roughness indirectly revealed that the change in the WO₃ film composition and crystalline structure was due to the increase in the annealing temperature, which is consistent with the results of XRD.

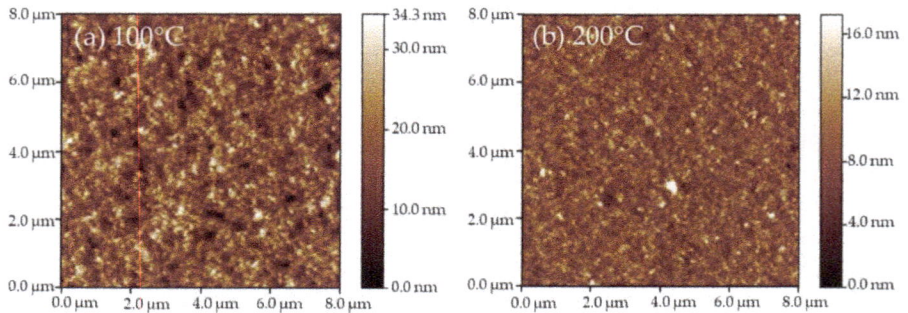

Figure 2. *Cont.*

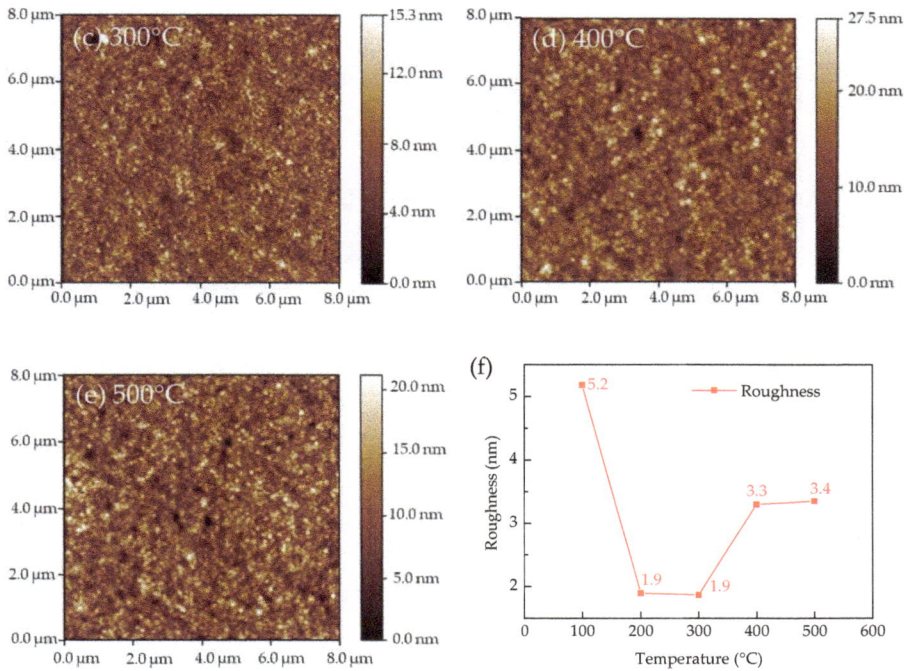

Figure 2. The atomic force microscope (AFM) images 8000 nm × 8000 nm) and the roughness of WO$_3$ films. (**a**) 100 °C; (**b**) 200 °C; (**c**) 300 °C; (**d**) 400 °C; (**e**) 500 °C; (**f**) the roughness of WO$_3$ films, which are read by the support software of AFM.

The band gap of WO$_3$ film can be measured and analyzed by an ultraviolet spectrophotometer. The optical band gap is distinguished from the band gap measured by other methods. According to Equation (1), the optical band gap can be calculated [23].

$$\alpha h\upsilon = A(h\upsilon - E_g)^n \tag{1}$$

where α is the absorption coefficient, which can be measured by the ultraviolet spectrophotometer; h is the Planck constant; υ is the light frequency; A is a proportionality constant; E_g is the optical band gap; and n is a number which is $1/2$ for the direct band gap semiconductor and 2 for the indirect band gap semiconductor. In this work, n is 2 because the WO$_3$ was an indirect band gap semiconductor.

To further investigate the electrochromic effects on the optical band gap of WO$_3$ film, the optical band gap of WO$_3$ film in a bleached state and colored state were analyzed. Electrochromism involves an electrochemical reaction, as shown in Equation (2) [24]:

$$WO_3(\text{colorless}) + xLi^+ + xe^- \leftrightarrow Li_xWO_3(\text{blue}) \tag{2}$$

At its bleached state, the WO$_3$ film is colorless. When both Li$^+$ and the electron are injected into the WO$_3$ film under an applied voltage, the bleached state of WO$_3$ turns into a colored state due to the generation of blue Li$_x$WO$_3$.

Figure 3a–e illustrates the curves of $(\alpha h\upsilon)^{1/2}$ versus the photon energy $h\upsilon$, which are calculated using the transmission spectra of the WO$_3$ films in the colored state and the bleached state. E_g can be extracted through the onset of the optical transitions of the WO$_3$ films near the band edge, which is equal to the value of the fitting line intercepts. Figure 3f shows a comparison of the optical band gap value of the WO$_3$ film that were annealed at different temperatures and electrochromic state (colored and bleached) and it indicates that the E_g of bleached WO$_3$ films decreases from 3.58 eV to 3.3 eV as the annealing temperature increases. Similarly, the E_g of the colored WO$_3$ film tends to decrease with an increased annealing temperature. In addition, the E_g of all the colored WO$_3$ films was less than that of their respective bleached WO$_3$ films.

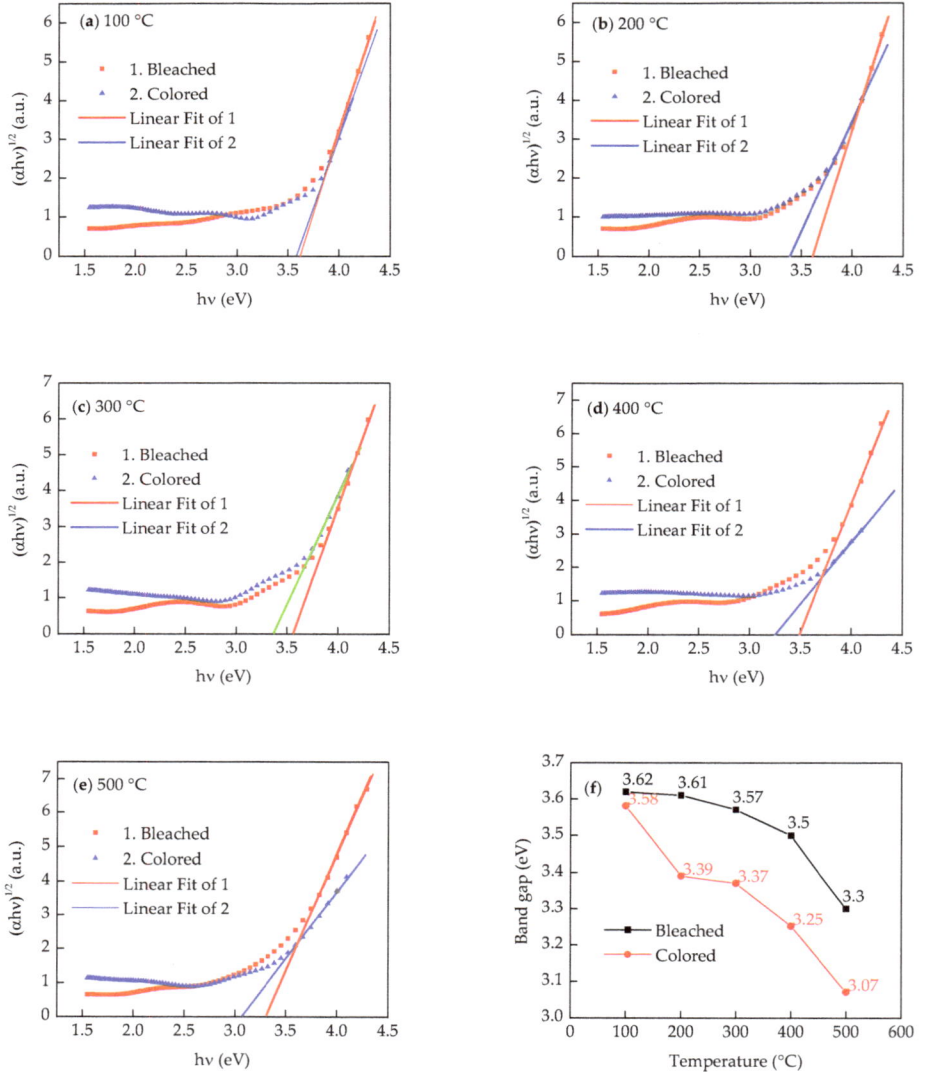

Figure 3. Optical band gap energy of WO$_3$ films in a colored state and bleached state. (**a**) 100 °C; (**b**) 200 °C; (**c**) 300 °C; (**d**) 400 °C; (**e**) 500 °C; and (**f**) a comparison of optical band gap energy of WO$_3$ films annealed at different temperature and electrochromic state (colored and bleached).

As for E_g, which decreased when the annealing temperature increased, a reasonable explanation was that as the annealing temperature increased, the oxygen vacancies increased, which may have provided free electrons and enhanced the conductivity of the WO_3 films [25].

To further investigate the relationship between E_g, conductivity, and electrochromic response time, an electrochromic test was conducted. Figure 4a,b illustrates the current density of the different WO_3 films and the change of transmittance (at 600 nm) under ± 2.5 V voltage, respectively. The peak current density of these films in the coloring process increased when the annealing temperature increased (an increase from 2.6 mA/cm^2 at 100 °C to 16.1 mA/cm^2 at 500 °C). This indicated that the conductivity enhanced with the increase in the annealing temperature. Similarly, the peak current density of these films in the bleaching process shows a similar change (increase from 11.0 mA/cm^2 at 100 °C to 22.2 mA/cm^2 at 500 °C). These were attributed to the decrease of E_g and the increase of free electrons. In addition, Figure 4a illustrates that the peak current density of the bleaching process was larger than that of coloring process, which results from the good conductivity of Li_xWO_3 [26]. This is related to the decrease of E_g after WO_3 film coloring.

Figure 4b illustrates an intuitive change of transmittance response curves. The response time is defined by the time corresponding to 90% of the total transmittance change. Figure 5 shows a specific comparison of the response time in the electrochromic test. The curve of the bleaching response time in Figure 5 shows that the bleaching response time increases from 1.2 s to 22.7 s, when the annealing temperature increased. In the bleaching process, the applied voltage drop is mainly across the electrolyte and the Li_xWO_3 layer. The extraction of Li$^+$ depends largely on the voltage across the Li_xWO_3 layer [27]. The E_g of the WO_3 film at the colored state reduced with the increase in the annealing temperature, which was attributed to the increase in the number of free electrons. In other words, the conductivity enhanced with the increase in annealing temperature. Therefore, the voltage across the Li_xWO_3 layer reduced with the increase in annealing temperature, which resulted in the increase of the bleaching response time. However, there was no similar trend in the coloring response time. The influence factors are not only the conductivity of the WO_3 film, but also the interface barrier of electrolyte-film [28]. The coloring response time increased when the WO_3 film changed from amorphous into crystalline, which resulted from the decrease in the voltage drop at the WO_3 layers, due to the increase in the conductivity. The band gap mainly influenced the transmission of the electrons, but the transmission of Li$^+$ depended more on the structure of films (such as crystallinity, morphology, etc.) [29].

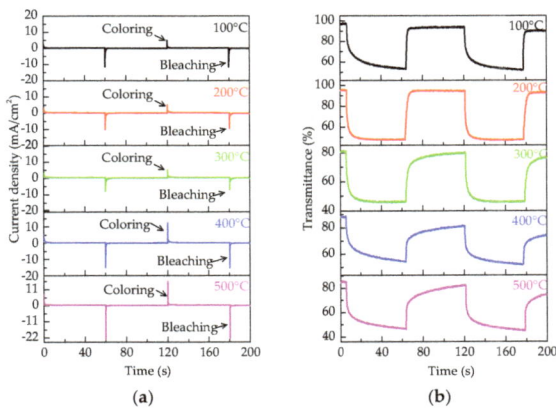

Figure 4. (**a**) Current change of WO_3 films at different annealing temperature. The applied voltage was ± 2.5 V and the WO_3 films were placed in the cathode; (**b**) change of transmittance (at 600 nm) of WO_3 films at different annealing temperature.

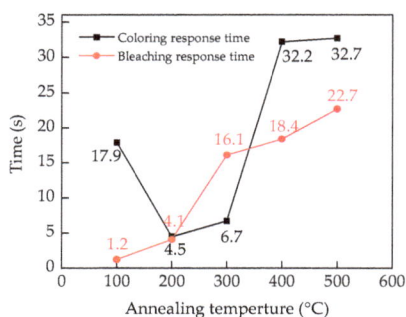

Figure 5. The curves of coloring and bleaching response time versus annealing temperature. The time corresponding to 90% of the total transmittance change is defined as the electrochromic response time.

4. Conclusions

The effects of the annealing temperature on the E_g of the WO_3 films were investigated. When the annealing temperature was higher than 400 °C, the crystalline structure of the WO_3 film changed from amorphous to monoclinic (400 °C), and then to cubic (500 °C). The E_g of the WO_3 films decreased from 3.62 eV to 3.30 eV when the annealing temperature was increased. In addition, the E_g of the colored WO_3 films was less than that of the bleached WO_3 films. The relationship between the E_g, conductivity, and electrochromic response time of the WO_3 film with different annealing temperatures demonstrates that the conductivity of the WO_3 film enhanced with the decrease in E_g, while the high conductivity increased the electrochromic response time.

Author Contributions: Conceptualization, G.Z.; Data curation, G.Z., K.L., W.Y., M.S. and X.L.; Formal analysis, G.Z., K.L., W.Y., R.T. and X.L.; Funding acquisition, H.N. and R.Y.; Investigation, X.Z. and M.S.; Methodology, X.Z., R.T. and X.L.; Project administration, H.N., R.Y. and J.P.; Supervision, H.N.; Writing—original draft, G.Z.; Writing—review & editing, K.L., H.N., R.T., X.L., R.Y. and J.P.

Acknowledgments: This work was supported by National Natural Science Foundation of China (Grant.51771074, 51521002 and U1601651), National Key R&D Program of China (No.2016YFB0401504 and 2016YFF0203600), National Key Basic Research and Development Program of China (973 program, Grant No.2015CB655004) Founded by Ministry of Science and Technology (MOST), Guangdong Natural Science Foundation (No.2016A030313459 and 2017A030310028), Guangdong Science and Technology Project (No.2016B090907001, 2016A040403037, 2016B090906002, 2017B090907016 and 2017A050503002), Guangzhou Science and Technology Project (201804020033).

Conflicts of Interest: The authors declare no conflicts of interest.

References

1. Hill, J.C.; Choi, K.S. Effect of electrolytes on the selectivity and stability of n-type WO_3 photoelectrodes for use in solar water oxidation. *J. Phys. Chem. C* **2012**, *116*, 7612–7620. [CrossRef]
2. Leidinger, M.; Huotari, J.; Sauerwald, T.; Lappalainen, J.; Schütze, A. Selective detection of naphthalene with nanostructured WO_3 gas sensors prepared by pulsed laser deposition. *J. Sens. Sens. Syst.* **2016**, *5*, 147–156. [CrossRef]
3. Sotelo-Vazquez, C.; Quesada-Cabrera, R.; Ling, M.; Scanlon, D.O.; Kafizas, A.; Thakur, P.K.; Lee, T.L.; Taylor, A.; Watson, G.W.; Palgrave, R.G. Photocatalysis: Evidence and effect of photogenerated charge transfer for enhanced photocatalysis in WO_3/TiO_2 heterojunction films: A computational and experimental study. *Adv. Funct. Mater.* **2017**, *27*, 1605413. [CrossRef]
4. Hara, K.; Zhao, Z.G.; Cui, Y.; Miyauchi, M.; Miyashita, M.; Mori, S. Nanocrystalline electrodes based on nanoporous-walled WO_3 nanotubes for organic-dye-sensitized solar cells. *Langmuir* **2011**, *27*, 12730–12736. [CrossRef] [PubMed]
5. Wang, F.; Valentin, C.D.; Pacchioni, G. Doping of WO_3 for photocatalytic water splitting: Hints from density functional theory. *J. Phys. Chem. C* **2012**, *116*, 8901–8909. [CrossRef]

6. Cai, G.; Cui, M.; Kumar, V.; Darmawan, P.; Wang, J.; Wang, X.; Eh, L.S.; Qian, K.; Lee, P.S. Ultra large optical modulation of electrochromic porous WO_3 film and the local monitoring of redox activity. *Chem. Sci.* **2016**, *7*, 1373–1382. [CrossRef] [PubMed]
7. Yang, W.; Runnerstrom, E.L.; Milliron, D.J. Switchable materials for smart windows. *Annu. Rev. Chem. Biomol. Eng.* **2016**, *7*, 283–304.
8. Granqvist, C.G. Electrochromic tungsten oxide films: Review of progress 1993–1998. *Sol. Energy Mat. Sol. Cells* **2000**, *60*, 201–262. [CrossRef]
9. Lemire, C.; Lollman, D.B.B.; Mohammad, A.A.; Gillet, E.; Aguir, K. Reactive R.F. magnetron sputtering deposition of WO_3 thin films. *Sens. Actuators B Chem.* **2002**, *84*, 43–48. [CrossRef]
10. Davazoglou, D.; Leveque, G.; Donnadieu, A. Study on the optical and electrochromic properties of polycrystalline WO_3 thin films prepared by CVD. *Sol. Energy Mater.* **1988**, *17*, 379–390. [CrossRef]
11. Reichman, B.; Bard, A.J. The Electrochromic process at WO_3 electrodes prepared by vacuum evaporation and anodic oxidation of W. *J. Electrochem. Soc.* **2016**, *126*, 583–591. [CrossRef]
12. Zayim, E.O. Optical and electrochromic properties of sol-gel made anti-reflective WO_3-TiO_2 films. *Sol. Energy Mate. Sol. Cells* **2005**, *87*, 695–703. [CrossRef]
13. Wojcik, P.; Cruz, A.; Santos, L.; Pereira, L.; Martins, R.; Fortunato, E. Microstructure control of dual-phase inkjet-printed a-WO_3/TiO_2/WO_X films for high-performance electrochromic applications. *J. Mater. Chem.* **2012**, *22*, 13268–13278. [CrossRef]
14. Santos, L.; Wojcik, P.; Pinto, J.V.; Elangovan, E.; Viegas, J.; Pereira, L.; Martins, R.; Fortunato, E. Structure and morphologic influence of WO_3 nanoparticles on the electrochromic performance of dual-phase a-WO_3/WO_3 inkjet printed films. *Adv. Electron. Mater.* **2015**, *1*, 1–2. [CrossRef]
15. Mukherjee, R.; Sahay, P.P. Effect of precursors on the microstructural, optical, electrical and electrochromic properties of WO_3 nanocrystalline thin films. *J. Mater. Sci-Mater. Electron.* **2015**, *26*, 1–13. [CrossRef]
16. Chai, Y.N.; Razak, K.A.; Lockman, Z. Effect of annealing on acid-treated $WO_3 \cdot H_2O$ nanoplates and their electrochromic properties. *Electrochimica Acta* **2015**, *178*, 673–681.
17. Gonzalez-Borrero, P.P.; Sato, F.; Medina, A.N.; Baesso, M.L.; Bento, A.C.; Baldissera, G.; Persson, C.; Niklasson, G.A.; Granqvist, C.G.; Silva, A.J.D. Optical band-gap determination of nanostructured WO_3 film. *Appl. Phys. Lett.* **2010**, *96*, 201. [CrossRef]
18. Vemuri, R.S.; Engelhard, M.H.; Ramana, C.V. Correlation between surface chemistry, density, and band gap in nanocrystalline WO_3 thin films. *ACS Appl.Mater. Interfaces* **2012**, *4*, 1371–1377. [CrossRef] [PubMed]
19. Min, H.K.; Choi, H.W.; Kim, K.H. Thickness dependence of WO_{3-x} thin films for electrochromic device application. *Mol. Cryst. Liquid Cryst.* **2014**, *598*, 54–61.
20. Arfaoui, A.; Ouni, B.; Touihri, S.; Mannoubi, T. Investigation into the optoelectrical properties of tungsten oxide thin films annealed in an oxygen air. *Mater. Res. Bull.* **2014**, *60*, 719–729. [CrossRef]
21. Szilágyi, I.M.; Santala, E.; Heikkilä, M.; Kemell, M.; Nikitin, T.; Khriachtchev, L.; Räsänen, M.; Ritala, M.; Leskelä, M. Thermal study on electrospun polyvinylpyrrolidone/ammonium metatungstate nanofibers: Optimising the annealing conditions for obtaining WO_3 nanofibers. *J. Therm. Anal. Calorim.* **2011**, *105*, 73. [CrossRef]
22. Badilescu, S.; Ashrit, P.V. Study of sol-gel prepared nanostructured WO_3 thin films and composites for electrochromic applications. *Solid State Ion.* **2003**, *158*, 187–197. [CrossRef]
23. Zou, Y.S.; Zhang, Y.C.; Lou, D.; Wang, H.P.; Gu, L.; Dong, Y.H.; Dou, K.; Song, X.F.; Zeng, H.B. Structural and optical properties of WO_3 films deposited by pulsed laser deposition. *J. Alloy Compd.* **2014**, *583*, 465–470. [CrossRef]
24. Haro-Poniatowski, E.; Jouanne, M.; Morhange, J.F.; Julien, C.; Diamant, R.; Fernández-Guasti, M.; Fuentes, G.A.; Alonso, J.C. Micro-raman characterization of WO_3 and MoO_3 thin films obtained by pulsed laser irradiation. *Appl. Surf. Sci.* **1998**, *127–129*, 674–678. [CrossRef]
25. Chai, Y.; Tam, C.W.; Beh, K.P.; Yam, F.K.; Hassan, Z. Effects of thermal treatment on the anodic growth of tungsten oxide films. *Thin Solid Films* **2015**, *588*, 44–49. [CrossRef]
26. Badot, J.C.; Beluze, L.; Dubrunfaut, O. Particle size effect on the electronic conductivity of electroactive $Li_xWO_3 \cdot H_2O$ powders: A study from 10^3 to 10^{10} Hz. *J. Phys. Chem. C* **2009**, *112*, 14549–14554. [CrossRef]
27. Faughnan, B.W.; Crandall, R.S.; Lampert, M.A. Model for the bleaching of WO_3 electrochromic films by an electric field. *Appl. Phys. Lett.* **1975**, *27*, 275–277. [CrossRef]

28. Srivastava, A.K.; Deepa, M.; Singh, S.; Kishore, R.; Agnihotry, S.A. Microstructural and electrochromic characteristics of electrodeposited and annealed WO_3 films. *Solid State Ion.* **2005**, *176*, 1161–1168. [CrossRef]

29. Koo, B.R.; Ahn, H.J. Fast-switching electrochromic properties of mesoporous WO_3 films with oxygen vacancy defects. *Nanoscale* **2017**, *9*, 17788–17793. [CrossRef] [PubMed]

micromachines

MDPI

Article

An Improved Large Signal Model for 0.1 *μ*m AlGaN/GaN High Electron Mobility Transistors (HEMTs) Process and Its Applications in Practical Monolithic Microwave Integrated Circuit (MMIC) Design in W band

Junfeng Li [1], Shuman Mao [1], Yuehang Xu [1,*], Xiaodong Zhao [1], Weibo Wang [2], Fangjing Guo [2], Qingfeng Zhang [1], Yunqiu Wu [1], Bing Zhang [2], Tangsheng Chen [2], Bo Yan [1], Ruimin Xu [1] and Yanrong Li [1]

[1] School of Electronic Science and Engineering (National Exemplary School of Microelectronics), University of Electronic Science and Technology of China, Chengdu 611731, China; lijunfeng_EE@163.com (J.L.); maoshuman@163.com (S.M.); zhaoxiaodong@std.uestc.edu.cn (X.Z.); a2277462594@163.com (Q.Z.); yqwu@uestc.edu.cn (Y.W.); yanbo@ee.uestc.edu.cn (B.Y.); rmxu@uestc.edu.cn (R.X.); yrli@uestc.edu.cn (Y.L.)
[2] Nanjing Electronic Devices Institute, Nanjing 210016, China; bobommic@163.com (W.W.); fjiguo@163.com (F.G.); binzhang_cetc55@aliyun.com (B.Z.); chentsh@vip.sina.com (T.C.)
* Correspondence: yuehangxu@uestc.edu.cn

Received: 28 June 2018; Accepted: 8 August 2018; Published: 10 August 2018

check for updates

Abstract: An improved empirical large signal model for 0.1 μm AlGaN/GaN high electron mobility transistor (HEMT) process is proposed in this paper. The short channel effect including the drain induced barrier lowering (DIBL) effect and channel length modulation has been considered for the accurate description of DC characteristics. In-house AlGaN/GaN HEMTs with a gate-length of 0.1 μm and different dimensions have been employed to validate the accuracy of the large signal model. Good agreement has been achieved between the simulated and measured S parameters, I-V characteristics and large signal performance at 28 GHz. Furthermore, a monolithic microwave integrated circuit (MMIC) power amplifier from 92 GHz to 96 GHz has been designed for validation of the proposed model. Results show that the improved large signal model can be used up to W band.

Keywords: AlGaN/GaN HEMT; DIBL effect; channel length modulation; power amplifier; W band

1. Introduction

Wide band gap semiconductor Gallium Nitride (GaN) high electron mobility transistors (HEMTs) are excellent candidates in high frequency power electronics due to their unique advantages of higher breakdown voltage and higher output power density [1]. With the rapid development of process, the feature size of GaN HEMTs have been shrinking to less than 0.1 μm. GaN HEMTs with good performance for application in W band have been reported [2–5]. Also, over the past few years, several GaN HEMT based monolithic microwave integrated circuits (MMICs) up to W-band have been developed, due to their applications in high speed wireless communications or radar systems [6]. A GaN MMIC power amplifier at 91 GHz was reported to have 1.7 W output power that is associated with 11% power added efficiency [7]. A W-Band MMIC power amplifier with 3.46 W/mm output power density and 21% associated power added efficiency was then reported. The associated power gain is 13.7 dB. It offers a peak small signal gain of 16.7 dB over 90–97 GHz [2].

For applications of these devices in circuit design, compact nonlinear device modeling plays an important role in practical design. Recently, a few physical based compact models have sprung up due to their advantages in less fitting parameters and good accuracy up to the Ka band [8–11]. However, things will be different when the frequency is up to W band. Firstly, the parasitic effect will become obvious with the increasing of frequency and make the parameter extraction more difficult [12,13]. This problem can be solved by FW-EM (Full-wave electromagnetic) simulation [14]. Secondly, along with the reduction of feature size, the short channel effect becomes obvious. This phenomenon will in the end give rise to shift of threshold voltage. Thirdly, the gradual channel approximation (GCA) that is used in many kinds of physical based compact model [15,16] is no more effective as the channel length modulation is obvious in short channel devices. These effects will largely decrease the accuracy of physical based compact model. The empirical modeling method has been widely used due to their excellent performance in convergence and accuracy [17–22]. An effective validation of large signal model is validated by on-wafer load-pull measurement [23,24]. However, due to the complication of load-pull measurement, only one input/output impedance is validated. Nevertheless, more input/output impedances need to be validated for a large signal model in practical MMIC power amplifier design [25].

In this paper, the short channel effect, including the DIBL effect and channel length modulation, is studied. An improvement for the accuracy of the area near the pinch-off region in IV curve is performed based on an empirical modeling method as the GCA is no more effective in most physical based model. In-house AlGaN/GaN HEMTs with gate length of 0.1 µm is used for validation of the model. Performance, including S parameters, DC characteristics, and large signal characteristics at 28 GHz is validated by on-wafer measurement. Finally, a MMIC power amplifier is designed based on the proposed model for further validation.

This paper is organized as follows. In Section 2, the investigation on short channel effect is presented. The modeling method of it, which is based on an empirical method, is given in detail. In Section 3, the proposed large signal model is validated with two GaN HEMTs with different gate width. In Section 4, a MMIC power amplifier based on the large signal model in this work is designed for further validation of the model in W band. Finally, in Section 5, the conclusion of this work is presented.

2. Model Description

2.1. Short Channel Effects

Along with the decrease of gate length, the short channel effect, such as the drain induced barrier lowering (DIBL) effect will become obvious. The thickness of the barrier will not only be modulated by gate voltage, but also drain voltage. This will, in the end, lead to the drift of threshold voltage along with the drain voltage. This phenomenon can be easily captured in the static IV curve of 0.1 µm AlGaN/GaN HEMTs with different gate width in this work, which have been shown in Figure 1.

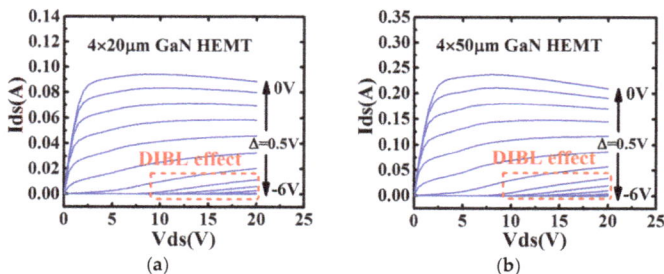

Figure 1. Drain induced barrier lowering (DIBL) effect in Static IV curves of 0.1 µm AlGaN/GaN high electron mobility transistor (HEMT) with different gate width: (**a**) 4 × 20 µm and (**b**) 4 × 50 µm.

It can be seen from Figure 1 that the DIBL effect will weaken the effect that is brought by gate voltage. The device will be turned from off-state to on-state with the rise of drain voltage. This phenomenon must be taken into consideration, especially for high efficiency power amplifier or switching applications.

In order to accurately describe the output performance of AlGaN/GaN HEMTs with short gate length in large signal modeling, the short channel effect, including the DIBL effect and channel length modulation, should be taken into consideration. An empirical method that is based on the Angelov model is employed for the devices in this work. As we know that the coefficients of the ψ polynomial in Angelov model, which is shown in Equation (1), mainly affect the accuracy of the region close to pinch-off state.

$$\psi = P_1 \times (V_{gs} - V_{pk1}) + P_2 \times \left(V_{gs} - V_{pk2}\right)^2 + P_3 \times \left(V_{gs} - V_{pk3}\right)^3 \tag{1}$$

where V_{gs} refers to the gate-source voltage. V_{pkn} ($n = 1, 2, 3$) are fitting parameters. P_n ($n = 1, 2, 3$) are fitting coefficients of the ψ polynomial.

To accurately model the DIBL effect, the drain-source voltage V_{ds} has been included in P_n ($n = 1, 2, 3$) to take the modulation effect of V_{ds} into consideration, as shown in Equation (2).

$$P_n = P_{n0} + (P_{n1} \times V_{ds} - P_{n0}) \times \tanh(\alpha P_{n2} \times V_{ds}) \quad (n = 1, 2, 3) \tag{2}$$

where P_{n0}, P_{n1}, P_{n2} and α are all fitting parameters.

The modification was validated by a comparison between simulation results and measured data. The comparison between the original Angelov model and modified one are shown in Figure 2. The gate-source voltage V_{gs} is from -6 V to -3 V and the drain source voltage V_{ds} is from 0 V to 20 V.

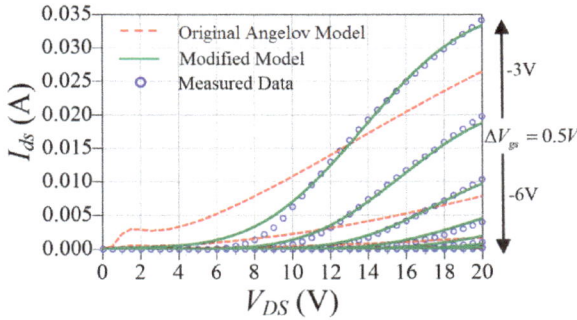

Figure 2. Comparison between simulated and measured results when V_{gs} is close to pinch-off voltage.

It is clear in Figure 2 that the original Angelov model cannot accurately describe the DC characteristics when V_{gs} is close to the pinch-off voltage. The DIBL effect can be successfully modeled by using proposed model.

Apart from the DIBL effect, the channel length modulation can also be captured in the static IV curves, as shown in Figure 3.

It clearly shows that the partial derivative of I_{ds} to V_{ds} is not equal to zero due to channel length modulation. The channel length effect is mainly induced by expanding of the depletion region towards the source. The effective channel is then shortened. This phenomenon is shown in Figure 4.

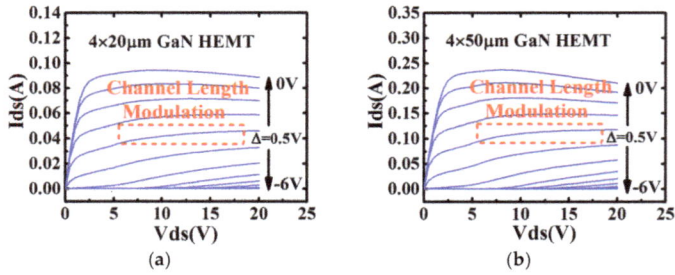

Figure 3. Channel length modulation effect in Static IV curves of 0.1 μm AlGaN/GaN HEMT with different gate width: (**a**) 4 × 20 μm and (**b**) 4 × 50 μm.

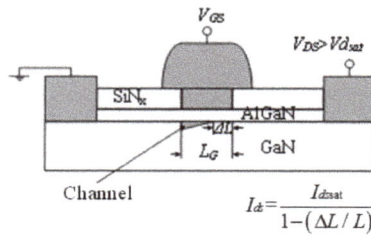

Figure 4. The schematic diagram of the short channel modulation effect.

2.2. Large Signal Model up to W Band

With the frequency up to W band, the RF dispersion will become more and more obvious due to the parasitic effects inside devices. A wide band small signal model [14], which has been proved to be able to cover the frequency band from 0.2–110 GHz, is employed in this work. The topology of the large signal model is shown in Figure 5.

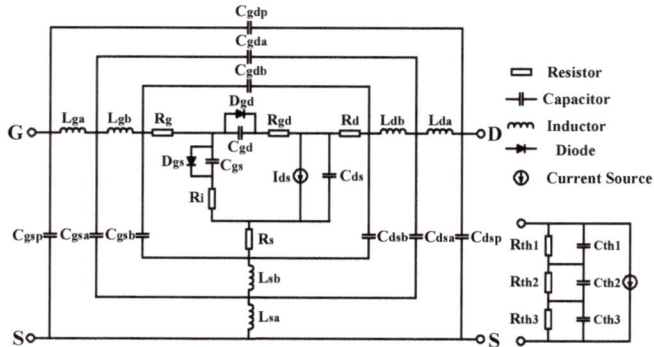

Figure 5. Topology of Large signal model up to W band.

The main part of the nonlinear current model as well as the capacitance model, including C_{gs} and C_{gd} mentioned in [21], is employed in this work. The improvement for accurate characterization of short channel effect, which is mentioned in the previous section, has also been included in the nonlinear current model. In order to accurately characterize the self-heating effect in AlGaN/GaN HEMT. The three-pole thermal network in [25] is used. Thermal resistances as well as the thermal capacitances are extracted by a method based on FEM simulation in ANSYS. The trapping effect is modeled by the equivalent voltage method in [26]. The scalability of the model parameters, including the I_{pk0}, R_{th}, and C_{th} has been realized with the method that is mentioned in [22] for practical monolithic microwave integrated circuit design. With the help of MATLAB coding, model parameters, except the coefficients in Equation (2), are all extracted with the method in [27]. In terms of parameters in Equation (2), they are all extracted by fitting the transfer characteristics curve with the least square method.

3. Model Validation

3.1. Small Signal Characterization

The large signal model was embedded into Keysight ADS (Advanced Design System) by a symbolically defined device (SDD) tool. Small signal characteristics of the devices are measured by cascade probe station (Summit 11000B, FormFactor, Livermore, CA, USA), which is shown in Figure 6. The vector network analyzer is Keysight N5247A (Keysight Technologies, Santa Rosa, CA, USA). The frequency extenders close to probes are used to achieve the S parameters ranging from 75 GHz to 110 GHz as the vector network analyzer can only reach up to 67 GHz.

Figure 6. On-wafer measurement system for small signal characteristics.

The proposed model was validated by 0.1 µm AlGaN/GaN HEMTs with different gate width. AlGaN/GaN HEMTs were all fabricated on a 4-inch SiC substrate. T-shape-gate technology was introduced to reduce the contact resistance. The f_T of the 0.1 µm GaN process is 90 GHz, while f_{max} is 220 GHz. The peak power density for a specific device can reach up to 3.46 W/mm. The photography of devices is shown in Figure 7.

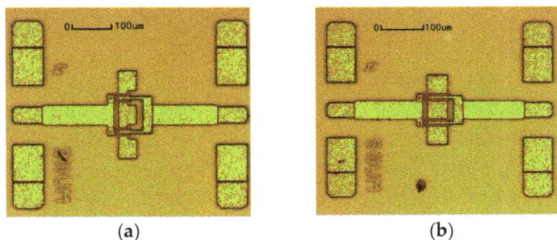

Figure 7. 0.1 µm AlGaN/GaN HEMTs: (a) 4×20 µm and (b) 4×50 µm.

The comparison of simulated and measured S parameters is shown in Figure 8. Results show that the proposed model can predict the small signal characteristics ranging from 0.2 GHz to 110 GHz for devices with different gate width and under different bias.

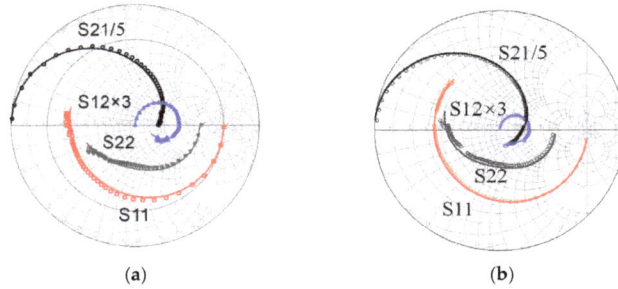

Figure 8. Comparison of simulated and measured S-parameters: (**a**) 4×20 μm at $Vgs = -2$ V, $Vds = 10$ V and (**b**) 4×50 μm at $Vgs = -1$ V, $Vds = 15$ V.

3.2. The Large Signal Model Validation

The DC characteristics for the proposed scalable large signal model was validated by different gate width, including 4×20 μm and 4×50 μm, as shown in Figure 9. The gate-source voltage V_{gs} is investigated from -6 V to 0 V, while the drain-source voltage V_{ds} is from 0 V to 20 V for these two devices.

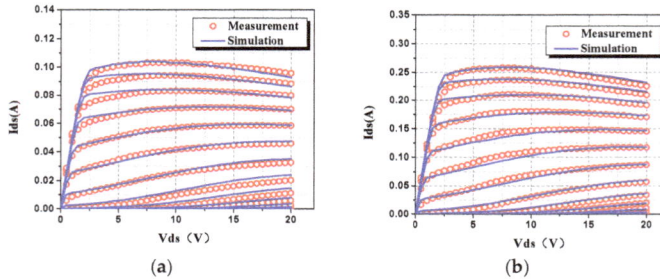

Figure 9. Comparison of simulated and measured DC characteristics of 0.1 μm AlGaN/GaN HEMTs: (**a**) 4×20 μm and (**b**) 4×50 μm.

Figure 9 shows that the DIBL effect is accurately characterized based on the improvement in Equation (2). The channel length modulation effect is also the same.

Due to the absent of W band load-pull system, the load pull performance at 28 GHz was used to validate the large signal model first, as shown in Figure 10. The system is on cascade probe station (Summit 12000, FormFactor, Livermore, CA, USA), the input signal generator is Agilent E8257D (Keysight Technologies, Santa Rosa, CA, USA), and the output power is detected by power meter Agilent N1912A (Keysight Technologies, Santa Rosa, CA, USA) and Vector Network Analyzer (Keysight Technologies, Santa Rosa, CA, USA).

The maximum output power load-pull measurement is performed. The bias is chosen at $V_{gs} = -2.6$ V, $V_{ds} = 15$ V, which is at deep class AB working state. The quiescent drain current is 82 mA at this bias. The optimum source and load resistance for the maximum output power are $Z_S = (13.44 + 12.41 \times j)$ Ω and $Z_L = (27.19 + 27.44 \times j)$ Ω. The power sweep was then performed based on the optimum resistance with the input power ranging from -4 dBm to 22 dBm. The comparison between the simulated and measured results, including output power (Pout), gain, and power added efficiency (PAE) are shown in Figure 11. Also, the influence that is brought by the DIBL effect has also been investigated in Figure 11. Results show that the DIBL effect will lead to the reduction of Pout, gain, and PAE. This can be explained by the variation of static bias point due to the DIBL effect.

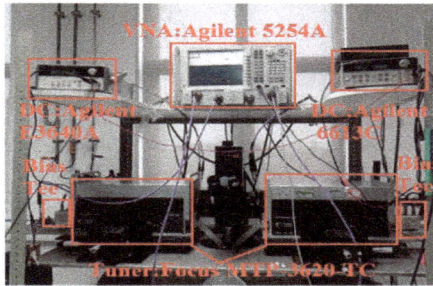

Figure 10. Photograph of on-wafer load–pull system setup.

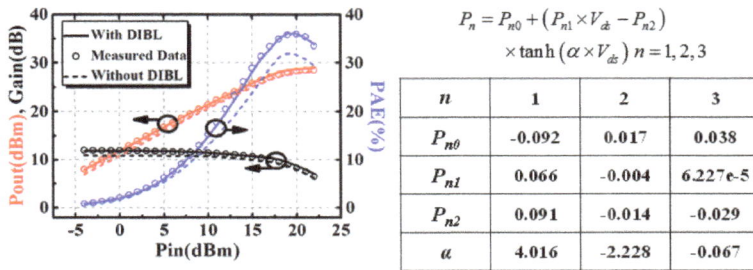

$$P_n = P_{n0} + \left(P_{n1} \times V_{ds} - P_{n2} \right)$$
$$\times \tanh \left(\alpha \times V_{ds} \right) \, n = 1, 2, 3$$

n	1	2	3
P_{n0}	-0.092	0.017	0.038
P_{n1}	0.066	-0.004	6.227e-5
P_{n2}	0.091	-0.014	-0.029
α	4.016	-2.228	-0.067

Figure 11. Investigation on the influence brought by DIBL effect on large signal performance.

The simulated and measured impedance charts achieved by maximum Pout and PAE load-pull measurement are presented in Figure 12.

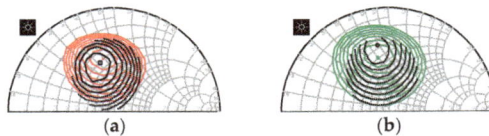

Figure 12. Comparison between simulated impedance chart and measured one: (**a**) maximum Pout and (**b**) maximum power added efficiency (PAE).

4. W Band MMIC Power Amplifier Design

For further validation of the proposed large signal model for applications in the W band, a MMIC power amplifier whose operation frequency is 92 GHz–96 GHz was designed. Based on the above large signal model, a W-band power amplifier is designed. Figure 13 presents the schematic of the W band amplifier.

The output stage used the planar spatial power combiner to realize the impedance transformation and combine the four-way power element. The millimeter wave GaN device is very easy to oscillation at low frequency due to the high gain. Multi-order RC network was used to improve the stability of the circuit. In order to enable the former stage to have enough power to drive the latter stage, the driving ratio of amplifier circuit is 1:2:4. Passive components include micro-strip line, MIM (Metal-insulator-Metal) capacitance, and resistor. All of the passive components were simulated by EM simulator in ADS. Figure 14 shows photograph of a W-band GaN MMIC amplifier.

(a)

(b)

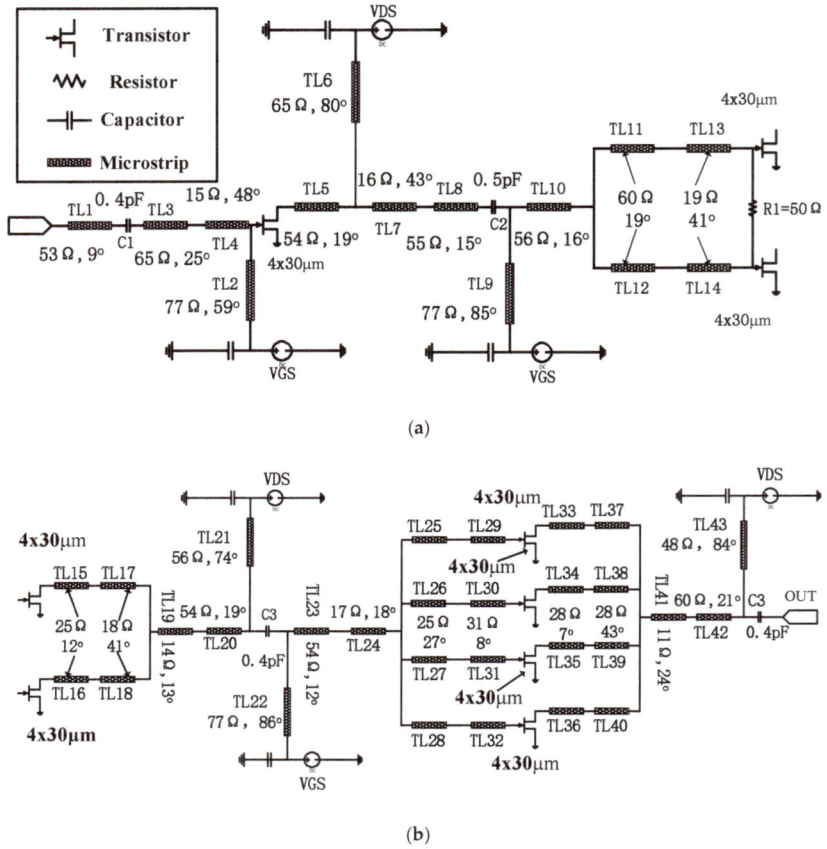

Figure 13. Schematic of W band amplifier: (**a**) Preceding stage and (**b**) Post stage.

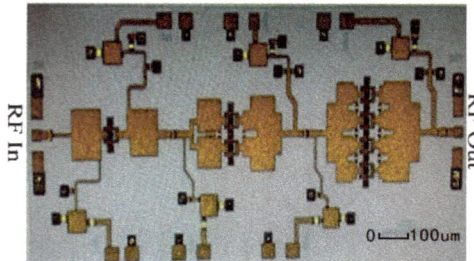

Figure 14. Photograph of a W-band Gallium Nitride (GaN) monolithic microwave integrated circuits (MMIC) amplifer.

The chip was loaded into a jig for measurement. The schematic of the measurement setup for large-signal measurements is shown in Figure 15. The large signal measurement was performed at room temperature. The commercial amplifier, frequency multiplier, and signal analyzer in Figure 15 are used to assistant the measurement. Other instruments including power meter (VDI Erickson, Virginia Diodes, Inc., Charlottesville, VA, USA), DC sources (Agilent E3633A and E3634A, Keysight Technologies, Santa Rosa, CA, USA), and attenuator (Rebes, Suzhou, China) were also employed. The amplifier is measured in CW (Continuous Wave) mode over 90 GHz–97 GHz frequency. The device was bias at $V_{ds} = 15$ V and $V_{gs} = -2$ V.

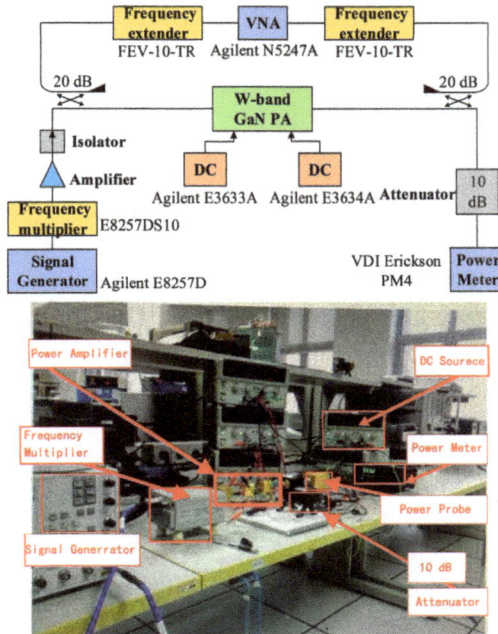

Figure 15. Photograph of the measurement setup for the W band MMIC power amplifier.

Figure 16 displays measured and simulated S-parameters of the MMIC amplifier. The difference in Figure 16 may come from the cavity and gold wire used for assisting the measurement. Their influence on frequency shift has not been taken into consideration during the MMIC design. However, this accuracy is sufficient for the application of practical circuit design. Figure 17 shows Gain, PAE, and output power. Over 90 GHz–97 GHz frequency range, the output power is greater than 1 W. The peak output power is 1.2 W. Except for 94 GHz and 98 GHz, the measured PAE was greater than 15%.

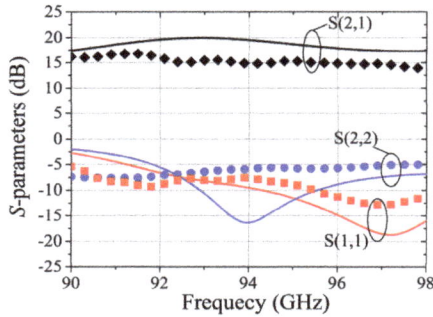

Figure 16. Measured (solid) and simulated (Symbol) S parameters of W band MMIC amplifier.

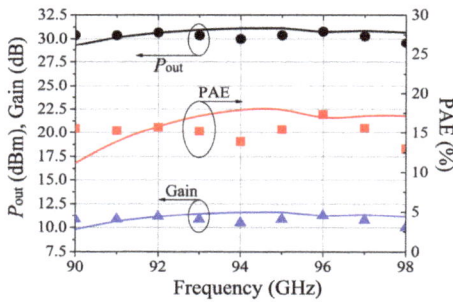

Figure 17. Measured (Symbol) and simulated (solid) large-signal characteristics of the W-band MMIC PA.

5. Conclusions

In this paper, an improved large signal model for AlGaN/GaN HEMT up to the W band is presented. The short channel effects including the DIBL effect and channel length modulation are added in the Angelov model. In-house AlGaN/GaN HEMTs with gate length of 0.1 μm are used for the validation of the model. A MMIC power amplifier is designed based on the proposed model for further validation. Results show that the large signal model can give good accuracy up to W band. The results of this paper can provide guidance to many other kinds FET (Field Effect Transistor) devices modeling in the W band. Also, they are useful for the improvement of the GaN process and also are helpful for the practical MMIC design in the W band.

Author Contributions: Investigation—J.L., S.M. and X.Z.; Methodology—Y.X.; Supervision—Y.X., B.Z., T.C., B.Y., R.X. and Y.L.; Validation—W.W., F.G., Q.Z. and Y.W.; Writing original draft—J.L. and S.M.; Writing review & editing—Y.X., J.L. and S.M. contributed equally to this work.

Funding: This research was funded by National Natural Science Foundation of China (Grant No. 61474020), the Fundamental Research Funds for the Central Universities (Grant No. ZYGX2016J036), and the National Key Project of Science and Technology.

Conflicts of Interest: The authors declare that there is no conflict of interests regarding the publication of this article.

References

1. Mishra, U.K.; Shen, L.; Kazior, T.E.; Wu, Y.F. Gan-based RF power devices and amplifiers. *Proc. IEEE.* **2008**, *96*, 287–305. [CrossRef]
2. Shaobing, W.; Jianfeng, G.; Weibo, W.; Junyun, Z. W-band MMIC PA with ultrahigh power density in 100-nm AlGaN/GaN technology. *IEEE Trans. Electron Devices* **2016**, *63*, 3882–3886. [CrossRef]
3. Wienecke, S.; Romanczyk, B.; Guidry, M.; Li, H.; Ahmadi, E.; Hestroffer, K.; Zheng, X.; Keller, S.; Mishra, U.K. N-polar gan cap mishemt with record power density exceeding 6.5 W/mm at 94 GHz. *IEEE Electron Device Lett.* **2017**, *38*, 359–362. [CrossRef]
4. Xing, W.; Liu, Z.; Ranjan, K.; Ng, G.I.; Palacios, T. Planar nanostrip-channel Al_2O_3/InAIN/GaN MISHEMTs on Si with improved linearity. *IEEE Electron Device Lett.* **2018**, *39*, 947–950. [CrossRef]
5. Romanczyk, B.; Wienecke, S.; Guidry, M.; Li, H.; Ahmadi, E.; Zheng, X.; Keller, S.; Mishra, U.K. Demonstration of constant 8 W/mm power density at 10, 30, and 94 GHz in state-of-the-art millimeter-wave N-polar GaN MISHEMTs. *IEEE Trans. Electron Devices* **2018**, *65*, 45–50. [CrossRef]
6. Niida, Y.; Kamada, Y.; Ohki, T.; Ozaki, S.; Makiyama, K.; Minoura, Y.; Okamoto, N.; Sato, M.; Joshin, K.; Watanabe, K. 3.6 W/mm high power density W-band InAlGaN/GaN HEMT MMIC power amplifier. In Proceedings of the 2016 IEEE Topical Conference on Power Amplifiers for Wireless and Radio Applications (PAWR), Austin, TX, USA, 24–27 January 2016; pp. 24–26.
7. Brown, A.; Brown, K.; Chen, J.; Hwang, K.C.; Kolias, N.; Scott, R. W-band GaN power amplifier MMICs. In Proceedings of the 2011 IEEE MTT-S International Microwave Symposium, Baltimore, MD, USA, 5–10 June 2011; pp. 1–4.
8. Cheng, X.; Wang, Y. A surface-potential-based compact model for AlGaN/GaN MODFETs. *IEEE Trans. Electron Devices* **2011**, *58*, 448–454. [CrossRef]
9. Khandelwal, S.; Chauhan, Y.S.; Fjeldly, T.A. Analytical modeling of surface-potential and intrinsic charges in AlGaN/GaN HEMT devices. *IEEE Trans. Electron Devices* **2012**, *59*, 2856–2860. [CrossRef]
10. Deng, W.; Huang, J.; Ma, X.; Liou, J.J. An explicit surface potential calculation and compact current model for AlGaN/GaN HEMTs. *IEEE Electron Device Lett.* **2015**, *36*, 108–110. [CrossRef]
11. Radhakrishna, U.; Choi, P.; Grajal, J.; Peh, L.S.; Palacios, T.; Antoniadis, D. Study of RF-circuit linearity performance of GaN HEMT technology using the MVSG compact device model. In Proceedings of the 2016 IEEE International Electron Devices Meeting (IEDM), San Francisco, CA, USA, 3–7 December 2016.
12. Dambrine, G.; Cappy, A.; Heliodore, F.; Playez, E. A new method for determining the FET small-signal equivalent circuit. *IEEE Trans. Microw. Theory Tech.* **1988**, *36*, 1151–1159. [CrossRef]
13. Jarndal, A.; Kompa, G. A new small-signal modeling approach applied to GaN devices. *IEEE Trans. Microw. Theory Tech.* **2005**, *53*, 3440–3448. [CrossRef]
14. Jia, Y.; Xu, Y.; Xu, R.; Li, Y. An accurate parasitic parameters extraction method based on FW-EM for AlGaN/GaN HEMT up to 110 GHz. *Int. J. Numer. Model. Electron. Netw. Devices Fields* **2018**, *31*, e2270. [CrossRef]
15. Wen, Z.; Xu, Y.; Chen, Y.; Tao, H.; Ren, C.; Lu, H.; Wang, Z.; Zheng, W.; Zhang, B.; Chen, T.; et al. A quasi-physical compact large-signal model for AlGaN/GaN HEMTs. *IEEE Trans. Microw. Theory Tech.* **2017**, *65*, 5113–5122. [CrossRef]
16. Wu, Q.; Xu, Y.; Chen, Y.; Wang, Y.; Fu, W.; Yan, B.; Xu, R. A scalable multiharmonic surface-potential model of AlGaN/GaN HEMTs. *IEEE Trans. Microw. Theory Tech.* **2018**, *66*, 1192–1200. [CrossRef]
17. Crupi, G.; Xiao, D.; Schreurs, D.M.M.P.; Limiti, E.; Caddemi, A.; Raedt, W.D.; Germain, M. Accurate multibias equivalent-circuit extraction for GaN HEMTs. *IEEE Trans. Microw.Theory Tech.* **2006**, *54*, 3616–3622. [CrossRef]
18. Jardel, O.; Groote, F.D.; Reveyrand, T.; Jacquet, J.C.; Charbonniaud, C.; Teyssier, J.P.; Floriot, D.; Quere, R. An electrothermal model for AlGaN/GaN power HEMTs including trapping effects to improve large-signal simulation results on high VSWR. *IEEE Trans. Microw. Theory Tech.* **2007**, *55*, 2660–2669. [CrossRef]
19. Liu, L.S.; Ma, J.G.; Ng, G.I. Electrothermal large-signal model of III–V FETs including frequency dispersion and charge conservation. *IEEE Trans. Microw. Theory Tech.* **2009**, *57*, 3106–3117.
20. Zhao, X.; Xu, Y.; Jia, Y.; Wu, Y.; Xu, R.; Li, J.; Hu, Z.; Wu, H.; Dai, W.; Cai, S. Temperature-dependent access resistances in large-signal modeling of millimeter-wave AlGaN/GaN HEMTs. *IEEE Trans. Microw. Theory Tech.* **2017**, *65*, 2271–2278. [CrossRef]

21. Wang, C.; Xu, Y.; Yu, X.; Ren, C.; Wang, Z.; Lu, H.; Chen, T.; Zhang, B.; Xu, R. An electrothermal model for empirical large-signal modeling of AlGaN/GaN HEMTs including self-heating and ambient temperature effects. *IEEE Trans. Microw. Theory Tech.* **2014**, *62*, 2878–2887. [CrossRef]

22. Xu, Y.; Wang, C.; Sun, H.; Wen, Z.; Wu, Y.; Xu, R.; Yu, X.; Ren, C.; Wang, Z.; Zhang, B.; et al. A scalable large-signal multiharmonic model of AlGaN/GaN HEMTs and its application in C-band high power amplifier MMIC. *IEEE Trans. Microw. Theory Tech.* **2017**, *65*, 2836–2846. [CrossRef]

23. Joshin, K.; Ozaki, S.; Ohki, T.; Okamoto, N.; Niida, Y.; Makiyama, K. Millimeter-wave GaN HEMT model with VDS dependence of CDS for power amplifier applications. In Proceedings of the 2014 Asia-Pacific Microwave Conference, Sendai, Japan, 4–7 November 2014; pp. 582–584.

24. Cutivet, A.; Altuntas, P.; Defrance, N.; Okada, E.; Avramovic, V.; Lesecq, M.; Hoel, V.; Jaeger, J.C.D.; Boone, F.; Maher, H. Large-signal modeling up to W-band of AlGaN/GaN based high-electron-mobility transistors. In Proceedings of the 2015 10th European Microwave Integrated Circuits Conference (EuMIC), Paris, France, 7–8 September 2015; pp. 93–96.

25. King, J.B.; Brazil, T.J. Nonlinear electrothermal GaN HEMT model applied to high-efficiency power amplifier design. *IEEE Trans. Microw. Theory Tech.* **2013**, *61*, 444–454. [CrossRef]

26. Yuk, K.S.; Branner, G.R.; McQuate, D.J. A wideband multiharmonic empirical large-signal model for high-power GaN HEMTs with self-heating and charge-trapping effects. *IEEE Trans. Microw. Theory Tech.* **2009**, *57*, 3322–3332. [CrossRef]

27. Wen, Z.; Xu, Y.; Wang, C.; Zhao, X.; Chen, Z.; Xu, R. A parameter extraction method for GaN HEMT empirical large-signal model including self-heating and trapping effects. *Int. J. Numer. Model. Electron. Netw. Devices Fields* **2017**, *30*, e2137. [CrossRef]

micromachines

MDPI

Article

AlGaN/GaN High Electron Mobility Transistors on Semi-Insulating Ammono-GaN Substrates with Regrown Ohmic Contacts

Wojciech Wojtasiak [1,*], **Marcin Góralczyk** [1], **Daniel Gryglewski** [1], **Marcin Zając** [2], **Robert Kucharski** [2], **Paweł Prystawko** [3,4], **Anna Piotrowska** [5,*], **Marek Ekielski** [5], **Eliana Kamińska** [5,*], **Andrzej Taube** [5] and **Marek Wzorek** [5]

[1] Institute of Radioelectronics and Multimedia Technology, Warsaw University of Technology, Nowowiejska 15/19, 00-662 Warsaw, Poland; M.Goralczyk@ire.pw.edu.pl (M.G.); dgrygle@ire.pw.edu.pl (D.G.)

[2] Ammono Lab, Institute of High Pressure Physics, Polish Academy of Sciences, Sokołowska 29/37, 01-142 Warsaw, Poland; zajac@ammono.com (M.Z.); kucharski@ammono.com (R.K.)

[3] TopGaN Ltd., Sokołowska 29/37, 01-142 Warsaw, Poland; pawel.prystawko@unipress.waw.pl

[4] Institute of High Pressure Physics, Polish Academy of Sciences, Sokołowska 29/37, 01-142 Warsaw, Poland

[5] Institute of Electron Technology, Al. Lotników 32/46, 02-668 Warsaw, Poland; ekielski@ite.waw.pl (M.E.); ataube@ite.waw.pl (A.T.); mwzorek@ite.waw.pl (M.W.)

* Correspondence: wwojtas@ire.pw.edu.pl (W.W.); ania@ite.waw.pl (A.P.); eliana@ite.waw.pl (E.K.); Tel.: +48-22-548-7940 (A.P.)

Received: 30 September 2018; Accepted: 21 October 2018; Published: 25 October 2018

check for updates

Abstract: AlGaN/GaN high electron mobility transistors on semi-insulating bulk ammonothermal GaN have been investigated. By application of regrown ohmic contacts, the problem with obtaining low resistance ohmic contacts to low-dislocation high electron mobility transistor (HEMT) structures was solved. The maximum output current was about 1 A/mm and contact resistances was in the range of 0.3–0.6 Ω·mm. Good microwave performance was obtained due to the absence of parasitic elements such as high access resistance.

Keywords: high electron mobility transistors; high electron mobility transistor (HEMT); AlGaN/GaN; ohmic contact; regrown contact; ammonothermal GaN; power amplifier

1. Introduction

There is a consensus in the nitride community that, although several GaN-based devices have already reached the market, their properties are still inferior with respect to predicted performance [1]. There are still many technological issues to be faced in order to fully exploit the enormous potential of these materials. The main limitations come from the lack of large area native bulk GaN substrates of reasonable cost and quality for homoepiaxial growth; some other issues concern device processing. Here, reliable low resistance homogeneous Ohmic contacts being fundamental building blocks of GaN devices are highly required.

In the quest to push the performance limit of AlGaN/GaN high electron mobility transistors (HEMTs), our work on advanced devices focuses on two main areas: development of epitaxial growth of HEMT structures on low defect density and high quality bulk ammonothermal semiinsulating GaN and fabrication of compatible nonalloyed ohmic contacts with subcontact n^+-In_xGa_{1-x}N:Si epilayer regrown by metal organic vapor phase epitaxy (MOVPE).

The AlGaN/GaN HEMT structures for high power applications are usually grown on silicon carbide substrates [2], but recently there has been significant progress in developing high quality

GaN substrates with low defect density using hydride vapor phase epitaxy and ammonothermal growth techniques [3–7]. In particular, truly bulk ammonothermal GaN substrates could be used for homoepitaxy of transistor active layers characterized by excellent crystal quality and low surface roughness due to low threading dislocation density at the order of 1×10^4 cm^{-2} [8] and negligible bow. This can lead to improved reliability, radiation hardness, high yield and repeatability of the parameters of the final devices [9–12].

In addition, due to the reliability and performance issues, the problem of self-heating and heat dissipation inside the epitaxial structure of GaN-based HEMTs is especially important for high power devices. While the thermal conductivity is higher for 4H-SiC than for bulk GaN, the heat flow inside typical GaN-based HEMT on the SiC substrate is significantly limited because of the presence of nucleation layers (e.g., AlN) between epilayers and SiC substrate. This effect is commonly called thermal boundary resistance (TBR) [13]. Dislocations at the interfaces have a large share in the TBR [14]. In the case of AlGaN/GaN structure grown on the bulk gallium nitride, the thermal resistance of GaN-based HEMT is only determined by the thermal conductivity of bulk GaN, and temperature rise under operating conditions can be comparable to that in devices on SiC substrates [15].

While preliminary results on some aspects of device technology were reported [16], AlGaN/GaN HEMTs on ammonothermal GaN substrates with satisfying DC and RF parameters have not yet been published. Moreover, difficulties in obtaining low resistivity ohmic contacts to HEMT structures made on a substrate with a lower dislocation density were reported [17]. In this work, we present AlGaN/GaN high electron mobility transistors on semi-insulating bulk ammonothermal GaN substrates with nonalloyed regrown ohmic contacts. By using metal organic vapor phase epitaxy (MOVPE)-regrown highly-doped n$^+$-In$_x$Ga$_{1-x}$N:Si layers, low resistivity ohmic contacts (R$_c$ ~0.3–0.6 Ω·mm) and high output current (1 A/mm) along with satisfying RF parameters are obtained.

2. Experimental Details

HEMT structures used in this study (see Figure 1a) were grown on a 1-inch c-plane, ~400 μm thick, semi-insulating ammonothermal bulk GaN (SI Ammono-GaN) substrates. The resistivity of semi-insulating Ammono-GaN is typically no less than 10^9 Ω·cm in paralel direction to the c-axis, as measured by frequency domain capacitive technique [18,19], and over 1×10^6 Ω·cm (above the measurement method range) in the perpendicular direction to c-axis, as determined by microwave methods [19,20].

High resistivity of SI Ammono-GaN substrates are obtained by compensation of residual oxygen, incorporated during ammonothermal growth, by deep acceptors i.e., transition metal ions or by Mg shallow acceptors. It is worth to noting that a low level of impurities (~2×10^{18} cm^{-3}) contributes to the high value of room temperature thermal conductivity of semi-insulating Ammono-GaN (κ ~230 W/mK) [21]. The HEMT structure was grown by MOVPE. It consists of 1 nm GaN-cap, 25 nm Al$_{0.26}$Ga$_{0.73}$N barrier layer, 0.8 nm AlN spacer, 0.7 μm unintentionally doped (UID) GaN and 1 μm GaN:C highly resistive buffer. As shown in Figure 1a a dotted line indicates the position of two-dimensional electron gas (2DEG) formed in the quantum well at AlN spacer/UID GaN interface.

The high resolution 2θ-ω and rocking curve X-ray diffraction (XRD) scans of epilayers (Figure 2a) prove the excellent crystal quality of epilayers grown on SI Ammono-GaN with FWHM = 0.007°. Atomic force microscopy (AFM) scans (5 μm × 5 μm) of the top of AlGaN/GaN HEMT structure shows an atomically smooth surface with a root mean square roughness of about 0.12–0.14 nm (Figure 2b). Electrical parameters of 2DEG were obtained by Hall effect and C-V measurements. Sheet resistivity (R$_{sh}$), sheet carrier concentration (n$_s$) and Hall mobility (μ) were 315 Ω/□, 1.64×10^{13} cm^{-2} and 1210 cm^2/Vs, respectively.

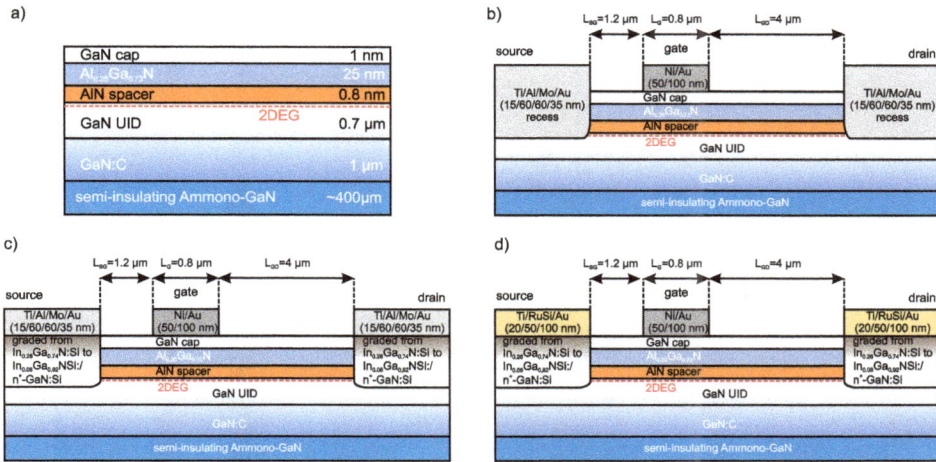

Figure 1. Cross-sectional schematics of the AlGaN/GaN-on-Ammono GaN high electron mobility transistors (HEMTs) under study: semiconductor device structure (**a**) and HEMT layouts with recessed Ti/Al/Mo/Au ohmic contact; (**b**) with subcontact n$^+$-In$_x$Ga$_{1-x}$N:Si regrown epilayer and Ti/Al/Mo/Au ohmic contact (**c**), and with subcontact n$^+$-In$_x$Ga$_{1-x}$N:Si regrown epilayer and Ti/RuSi/Au ohmic contact (**d**).

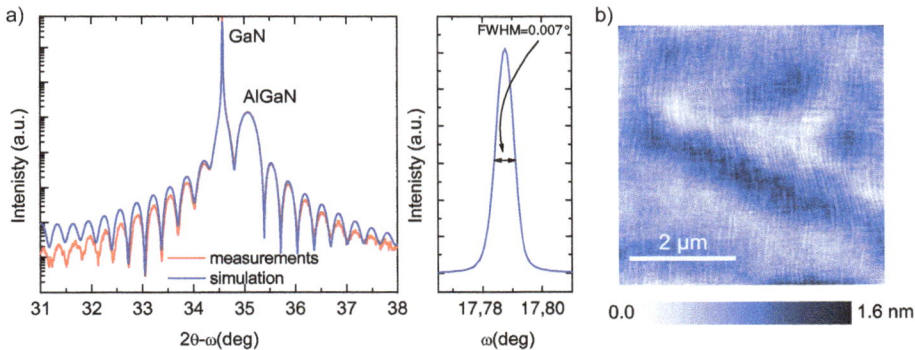

Figure 2. (**a**) 2θ-ω high resolution X-ray diffraction (XRD) scan and XRD rocking curve along with (**b**) AFM image of the surface of AlGaN/GaN HEMT structure on SI Ammono-GaN substrate.

The first step of HEMT processing was the deposition on the semiconductor device structure a double-layer SiO$_x$ (200 nm)/AlN (35 nm) mask and its patterning for the selective recess etching of AlGaN/GaN followed by selective MOVPE regrowth of n$^+$-In$_x$Ga$_{1-x}$N:Si/n$^+$-GaN:Si subcontact regions of source and drain. The SiO$_x$ film was deposited using plasma-enhanced chemical vapor deposition (PECVD) and AlN layer was grown by MOVPE. Mask patterning was performed by BCl$_3$/Ar and CHF$_3$/CF$_4$ plasma etching. The depth of recess etching was 20 nm below the AlGaN layer. The doping and thickness of subcontact regrown region was as follows: n$^+$-GaN:Si (Si: 1.7 × 10^{19} cm^{-3}–40 nm, 5 × 10^{19} cm^{-3}–7 nm) and graded (*x* from 8% to 26% at the top) n$^+$-In$_x$Ga$_{1-x}$N:Si (Si: 5 × 10^{19} cm^{-3}–10 nm). Schematic cross-section is presented in Figure 1c,d. To lower the surface barrier, doping of the first n$^+$-GaN:Si layer was kept below Mott concentration, while the next 7 nm n$^+$-GaN:Si layer was doped to the higher level of Si. For further lowering the surface barrier, n$^+$-In$_x$Ga$_{1-x}$N:Si graded layer was doped to 5 × 10^{19} cm^{-3} of Si and indium composition was chosen in such a way that with 26% of In, the Fermi level is pinned to the conduction band.

The AlN/SiO$_2$ mask was removed by soaking in hydrofluoric acid solution while regrown nitride films remained in contact regions. Figure 3 shows a cross-sectional transmission electron microscope (TEM) images of regrown GaN on top of a low-dislocation GaN homoepitaxial epilayer. High resolution imaging (HR-TEM) at Figure 3b reveals a smooth, dislocation-free n$^+$-GaN/UID GaN interface, the key attribute of GaN on GaN technology. The etched sidewalls are at a 56° angle to the c-plane (0001). The sidewall angle is close to the optimal 62° angle at which the density of dangling bonds on the etched surface is similar to a c-plane surface [22].

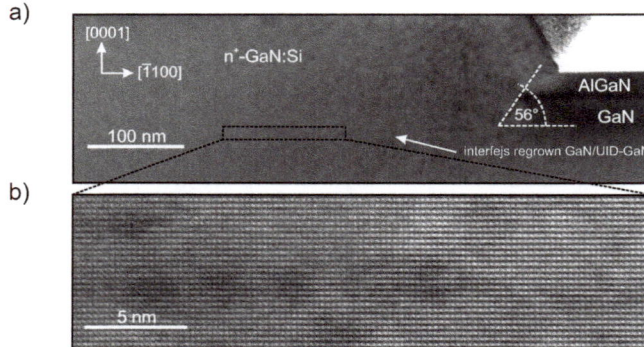

Figure 3. (**a**) TEM image of n$^+$-In$_x$Ga$_{1-x}$N:Si/n$^+$-GaN:Si subcontact region and (**b**) HR-TEM image of n$^+$-GaN:Si/UID GaN interface.

Next, the ohmic contact metallization was sputter-deposited and annealed. To compare the properties of alloyed and ohmic contacts, we studied the characteristics of conventional recessed Ti/Al/Mo/Au (15/60/60/35 nm) metallization annealed at 850 °C for 30 s in a nitrogen flow (Figure 1b) with contacts with subcontact n$^+$-In$_x$Ga$_{1-x}$N:Si regrown epilayer metallized using conventional Ti/Al/Mo/Au (15/60/60/35 nm) (Figure 1c) or thermally stable Ti/RuSi/Au (20/50/100 nm) metallizations(Figure 1d).

In the following, the isolation of adjacent devices was done by using two-step Al$^+$ ion implantation [23]. The implant consisted of Al ions at (1st step) energy of 800 keV, and dose 1.5×10^{13} cm^{-2} and (2nd step) at energy 300 keV, and dose 1×10^{13} cm^{-2}. The sheet resistivity of as-implanted isolation was 10^{11} Ω/□. With this technique, a sufficiently high vacancy density was obtained in the surface region down to 0.7 μm. To prevent the active regions from becoming implanted, a 3 μm thick photoresist mask was applied. Then, rectangular gate electrodes were an electron-beam deposited Ni/Au (50/100 nm) bilayer. Finally, the devices were passivated by 100 nm SiN$_x$ layer deposited by plasma-enhanced chemical vapor deposition. Finally, windows for contact pads were opened and pads were thickened by Au evaporation. The cross-section schematic of fabricated devices are shown in Figure 1b–d. The gate length (L$_G$) was 0.8 μm and gate width was 2 × 200 μm for two-finger devices. The source-gate (L$_{SG}$) and gate-drain distance (L$_{GD}$) were 1.2 μm and 4 μm, respectively.

3. Results and Discussion

3.1. Electrical Characterisation of Ohmic Contacts with Subcontact n$^+$-In$_x$Ga$_{1-x}$N:Si Regrown Epilayer to AlGaN/GaN Heterostrucutres on Semi-Insulating Ammono-GaN Substrates

As already mentioned in the Introduction, fabrication of low resistivity ohmic contacts to higher quality HEMT structures appears to be a difficult task. According to numerous studies of Ti/Al-based alloyed contacts to AlGaN/GaN 2DEG, the mechanism of ohmic contact formation is related to spiking through dislocations. Thus, the likely explanation of difficulties is the limited availability of dislocations in reduced defect density HEMTs [17,24,25].

One of the approaches to overcome this problem is to form a recess below 2DEG and form alloyed Ti/Al-based contact [26,27]. This method was reported successful for HEMTs with dislocation density above 10^6 cm^{-2} i.e., epistructures grown on SiC, Si or even HVPE (hydride vapour phase epitaxy) GaN substrates. In our case, the ohmics resistance of recessed alloyed Ti/Al/Mo/Au contacts to HEMT structures on SI Ammono-GaN are in the range of 0.8–1.1 Ω·mm. As a example, as shown in Figure 4a, the contact resistance and resistivity extracted from circular transmission line method (CTLM) measurements [28] for recessed Ti/Al/Mo/Au ohmic contacts were $R_C = 0.8$ Ω·mm and $\rho_c = 1.3 \times 10^{-5}$ Ω·cm^2.

Alloyed Ti/Al/Mo/Au ohmic makes contact with subcontact n$^+$-In$_x$Ga$_{1-x}$N:Si regrown epilayer show resistances from the range of 0.3–0.6 Ω·mm. For the HEMT structure used in this work, the contact resistance and resistivity were $R_C = 0.43$ Ω·mm and $\rho_c = 6.4 \times 10^{-6}$ Ω·cm^2, respectively, as extracted from CTLM measurements(Figure 4b). It is worth mentioning that measured contact resistance presents an upper limit of actual contact resistance as the measured value also includes the contribution of the n$^+$-GaN access region and regrown n$^+$-GaN-2DEG interface [29].

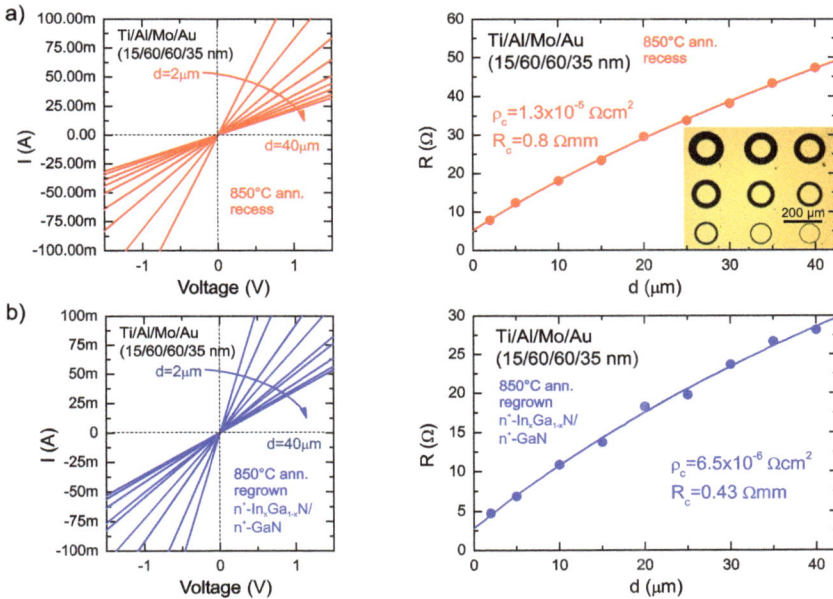

Figure 4. Four-point probe I-V plot of circular transmission line method (CTLM) patterns (image in inset) and measured resistance R vs. CTLM contact spacing, (the solid line is the result of fitting to experimental data) for (**a**) recessed and (**b**) Ti/Al/Mo/Au ohmic contacts with subcontact n$^+$-In$_x$Ga$_{1-x}$N:Si regrown epilayer annealed at 850 °C.

The non-alloyed ohmic contacts with Ti/Al/Mo/Au and Ti/RuSi/Au metallizations with subcontact n$^+$-In$_x$Ga$_{1-x}$N:Si regrown epilayer were mildly annealed at 400 °C for 1 min, in N$_2$ for promoting the adhesion. The comparison of current-characteristics and determination of contacts parameters using CTLM method is depicted in Figure 5. The contacts' resistance and reactivities obtained from the CTLM method were $R_C = 0.38 \pm 0.3$ ($\rho_c = 3.4 \times 10^{-6}$ Ω·cm^2) Ω·mm and $R_C = 0.43 \pm 0.8$ Ω·mm ($\rho_c = 5 \times 10^{-6}$ Ω·cm^2) for Ti/Al/Mo/Au and Ti/RuSi/Au contacts, respectively.

The use of regrown highly-doped In$_x$Ga$_{1-x}$N/GaN:Si makes it possible to create non-alloyed ohmic contacts to AlGaN/GaN heterostructures without high temperature annealing. Moreover, it allows for using thermally stable metal schemes, which allows for fabricating devices designed for high temperature applications. Sputter-deposited RuSi layers owing to amorphous microstructure [30,31] and high melting point, large work function and low resistivity are the material of choice for a diffusion barrier layer in metallization schemes They have already been proven reliable and thermally stable in GaN-based devices [32].

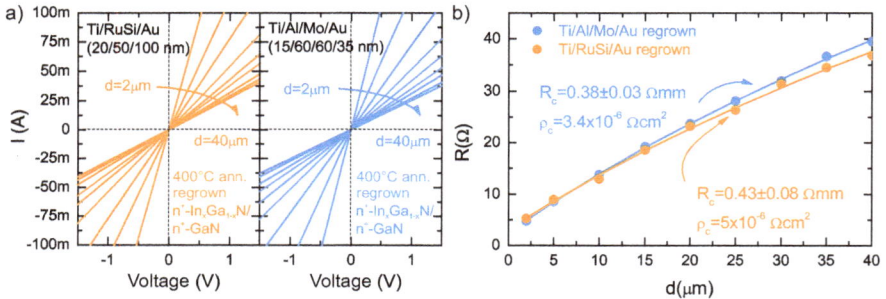

Figure 5. Four-point probe I-V plot of CTLM patterns (**a**) and measured resistance R vs. CTLM contact spacing (**b**) (the solid line is the result of fitting to experimental data) of Ti/Al/Mo/Au and Ti/RuSi/Au ohmic contacts with subcontact n$^+$-In$_x$Ga$_{1-x}$N:Si regrown epilayer annealed at 400 °C.

3.2. Electrical Characterization of AlGaN/GaN HEMTs on Semi-Insulating Ammono-GaN Substrates with Ohmic Contacts with Subcontact n$^+$-In$_x$Ga$_{1-x}$N:Si Regrown Epilayer

The output and transfer characteristics of the devices (with Ti/Al/Mo/Au ohmic contacts with subcontact n$^+$-In$_x$Ga$_{1-x}$N:Si regrown epilayer, annealed at 850 °C) are depicted in Figure 6a,b. The maximum drain current density for V$_{GS}$ = 2 V is about 1 A/mm. Extracted on-state resistance R$_{on}$ was 4.4 Ω·mm. The kink effect on the output characteristics is not observed. This effect is usually attributed to slow traps located in GaN buffer layer under gate region [33], which was described in previous reports on AlGaN/GaN HEMTs on ammonothermal bulk GaN [17]. The negative slope in the output characteristics for higher V$_{DS}$ and V$_{GS}$ values results from the self-heating [34,35]. The transconductance (g$_m$) is about 220 mS/mm and achieves maximum values for the expected range of operating points of transistor. The transfer characteristics show a clear pinch-off at V$_{GS}$ = −6 V and very good linear behaviour up to V$_{GS}$ = −2 V. The measured leakage current is about 0.1 mA/mm and can be attributed to the gate leakage current. We do not observe any additional, measurable leakage current through the buffer layers or the substrate.

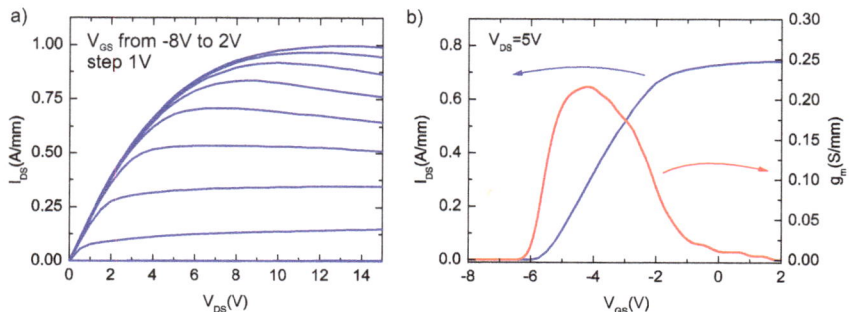

Figure 6. (**a**) output and (**b**) transfer characteristics of fabricated AlGaN/GaN HEMT on SI Ammono-GaN substrate with Ti/Al/Mo/Au ohmic contacts with subcontact n$^+$-In$_x$Ga$_{1-x}$N:Si regrown epilayer, annealed at 850 °C.

The frequency performance of transistors was also investigated. The *S*-parameters of fabricated transistors were measured over a 45 MHz to 24 GHz frequency range using on-wafer measurement station Cascade M150 with an Agilent N5242A network analyser (Keysight, Santa Rosa, CA, USA) and 50 Ω input and output impedance. Figure 7a shows RF characteristics such as current gain ($|h_{21}|$), maximum stable/available gain (MSG/MAG), unilateral gain (U) and $|S_{21}|$ gain at quiescent point V_{DS} = 28 V and I_{DQ} = 46 mA (115 mA/mm). The chosen operating point corresponds with condition to achieve maximum gain, at typical supply voltage used in standard GaN HEMT microwave circuits (V_{DS} = 28 V). The maximum frequency (f_{MAX}) and cut-off frequency (f_T) was 30 GHz and 21.1 GHz as obtained by linear extrapolation with −20 dB/dec slope of U (or MSG/MAG) and $|h_{21}|$, respectively. The f_T value (21.7 GHz) was also estimated using the Gummel method [36] as shown in Figure 7b. An $f_T \cdot L_g$ product of 16.8 GHz·μm was achieved. The $|S_{21}|$ gain attains 0 dB for frequency (f_s) of 22 GHz. The MAG and $|S_{21}|$ was 22.7 dB and 15.3 dB at 2 GHz and 19.8 dB and 12.7 dB at 4 GHz. It is worth noting that $|S_{21}|$ depends on the source and load impedances and the 50 Ω impedance of the measurement system is not optimal neither for maximum gain nor for maximal output power. Therefore, $|S_{21}|$ should not be used for direct comparison of the transistor structures. In order to estimate microwave properties and usability of the transistor, a small-signal model was extracted on the basis of the measured S-parameters. The measured and simulated input (S_{11}) and output (S_{22}) reflection coefficients are shown in Smith chart (Figure 7c) and forward (S_{21}) and reverse (S_{12}) transmission coefficients are plotted on the polar chart (Figure 7d). An equivalent circuit along with extracted model parameters are presented in Figure 7e. The microwave measurements indicate the lack of significant parasitic elements and confirm the high quality of fabricated HEMTs with ohmic contacts with subcontact n$^+$-In$_x$Ga$_{1-x}$N:Si regrown epilayer. The g_m value obtained from equivalent circuit was about 80.8 mS (200 mS/mm). As the slope of the $g_m(V_{GS})$ curve is steep, the DC g_m value of 200 mS/mm corresponds to I_{DS} value of about 120 mA/mm as can be deduced from Figure 6b. This value is close to the used I_{DQ} value, and confirms correspondence between the DC characteristics and the AC model.

In order to fully compare the Ti/Al/Mo/Au and Ti/RuSi/Au ohmic contacts with subcontact n$^+$-In$_x$Ga$_{1-x}$N:Si regrown epilayer, with reduced annealing temperature, another set of devices with each metallization were fabricated (on the parts of the same wafer), in a similar manner and with the same dimensions as described in the experimental section.

The comparison of output and transfer characteristics of AlGaN/GaN HEMTs with Ti/Al/Mo/Au and Ti/RuSi/Au ohmic contacts with subcontact n$^+$-In$_x$Ga$_{1-x}$N:Si regrown epilayer, annealed at 400 °C, are shown in Figure 8a–c. The transistors with Ti/RuSi/Au metallizations have a maximum output current of 756 mA/mm as compared to 736 mA/mm for the devices with Ti/Al/Mo/Au metallizations annealed at 400 °C. For both of the transistors, the pinch-off voltage was about −3.6 V, as is not dependent on used ohmic contact metallization schemes. It is worth noting that maximum output current is lower than described earlier; however, it is not connected with ohmic contact resistance but with parameters of 2DEG for AlGaN/GaN HEMT heterostructures used for fabrication of transistors, as suggested by lower pinch-off voltage values in those devices as compared to reported earlier in the text. The maximum transconductance value was 159 mS/mm at V_{GS} = −2.32 V for Ti/RuSi/Au devices and 167 mS/mm at V_{GS} = −1.79 V for the devices with Ti/Al/Mo/Au metallizations annealed at 400 °C.

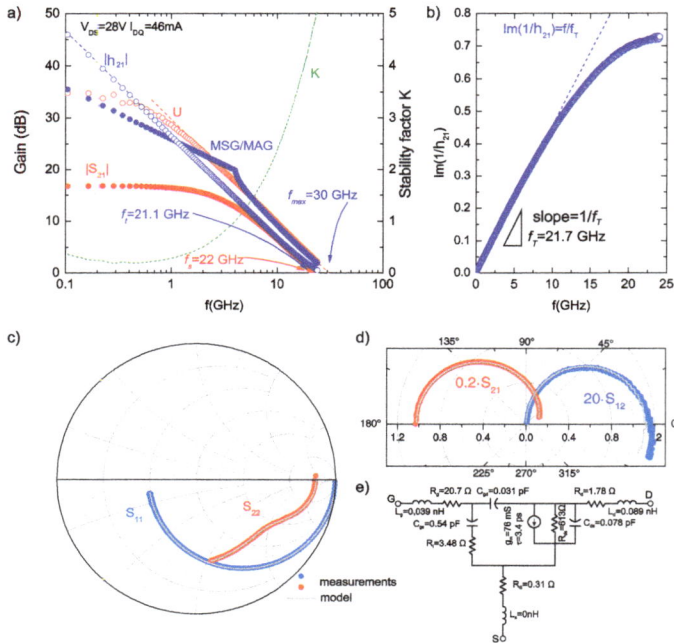

Figure 7. (**a**) high frequency characteristics of $|S_{21}|$, $|h_{21}|$, U and MSG/MAG of the fabricated AlGaN/GaN HEMT on SI Ammono-GaN substrates with Ti/Al/Mo/Au ohmic contacts with subcontact n^+-In_xGa_{1-x}N:Si regrown epilayer, annealed at 850 °C; (**b**) f_T determination using the Gummel method; (**c**) the Smith chart of S_{11} and S_{22} reflection coefficients; (**d**) section of polar plot of S_{21} and S_{12} transmission coefficients; (**e**) equivalent circuit with extracted parameters used for simulations.

Figure 8d,e shows the comparison of high-frequency characteristics (measured at quiescent point V_{DS} = 28 V) of AlGaN/GaN HEMTs on semi-insulating Ammono-GaN substrates with Ti/Al/Mo/Au and Ti/RuSi/Au ohmic contacts with subcontact n^+-In_xGa_{1-x}N:Si regrown epilayer, annealed at 400 °C. As can be seen, the transistors with both Ti/Al/Mo/Au and Ti/RuSi/Au exhibit a similar good high frequency performance and extracted dynamic parameters are very similar. Observed differences in those parameters do not deviate from typical values of parameter scattering for used technology. The maximum frequency (f_{MAX}) and cut-off frequency (f_T) was 28.4 GHz and 18.8 GHz and 28.8 GHz and 19.2 GHz for transistors with Ti/Al/Mo/Au and Ti/RuSi/Au ohmic contacts with subcontact n^+-In_xGa_{1-x}N:Si regrown epilayer, annealed at 400 °C, respectively, showing comparable values as for transistors with Ti/Al/Mo/Au ohmic contacts with subcontact n^+-In_xGa_{1-x}N:Si regrown epilayer, annealed at high temperature of 850 °C.

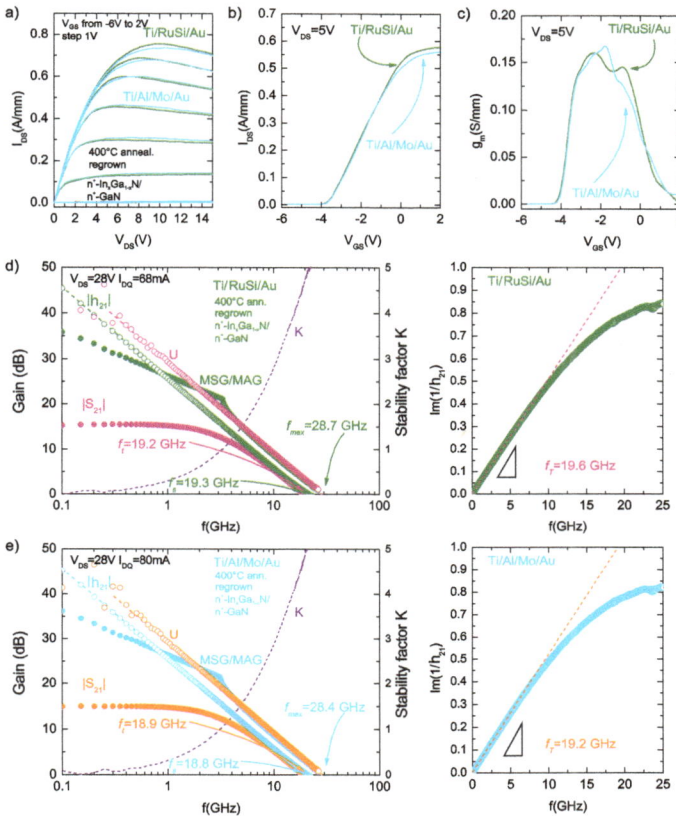

Figure 8. Comparison of (**a**) output and (**b,c**) transfer characteristics, (**d,e**) high frequency characteristics of $|S_{21}|$, $|h_{21}|$, U and MSG/MAG and f_T determination using the Gummel method for the fabricated AlGaN/GaN HEMT on SI Ammono-GaN substrates with Ti/Al/Mo/Au and Ti/RuSi/Au ohmic contacts with subcontact n^+-In$_x$Ga$_{1-x}$N:Si regrown epilayer, annealed at 400 °C.

3.3. Design and Fabrication of the on Microwave Power Amplifier Using AlGaN/GaN HEMTs on Semi-Insulating Ammono-GaN Substrates

To verify usability of the GaN HEMTs on Ammono GaN substrate in microwave designs, a class-AB power amplifier was designed using a small-signal approach based on a very popular Cripps method. This method enables a load impedance optimal for maximum output power level based on DC current-voltage characteristics and small-signal output impedance at a given transistor operating point to be determined [37]. The impedance condition recommended by Cripps suggests a series circuit, as the load, leaving aside the actual structure of transistor output circuit, which in the MESFET (metal semiconductor field effect transistor) and HEMT case has parallel circuit–parallel connection of C_{DS} and R_{DS} on the equivalent circuit (Figure 7e). Therefore, the definition of admittance condition in the plane of C_{DS} and R_{DS} elements seems natural. This approach is described in detail in [38,39]. In the assembly of the amplifier circuit, we use AlGaN/GaN HEMT on semi-insulating Ammono-GaN substrate with subcontact n^+-In$_x$Ga$_{1-x}$N:Si regrown epilayer and Ti/Al/Mo/Au ohmic contact annealed at 850 °C (representative characteristics of one of the devices from the same wafer are presented in Figures 6 and 7).

According to the aforementioned procedure, the amplifier was designed in a Keysight Advanced Design System (ADS) environment in the microstrip technique on Rogers RO4003C laminate (Chandler,

AZ, USA) with h = 0.02 and ϵ_r = 3.35. The output matching section was optimized to fulfill maximum power condition while the input matching circuit was optimized for minimum input return loss. The input network contains lossy elements to stabilize the amplifier. The assembly schematic and photography of the fabricated amplifier are presented in Figure 9, respectively. The use of a F&S BONDTEC 5632 bond machine (Braunau, Austria) ensures good control over the length and shape of bond wires. This is confirmed by the excellent compliance of simulations and measurements.

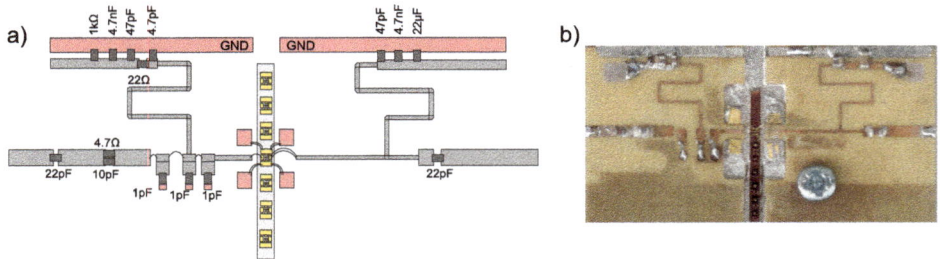

Figure 9. (**a**) assembly schematic and (**b**) photography of the fabricated amplifier.

The small-signal measurements were performed over a 2 GHz to 4 GHz frequency range and for input power level of −10 dBm. The simulations and measured results of the amplifier are compared in Figure 10.

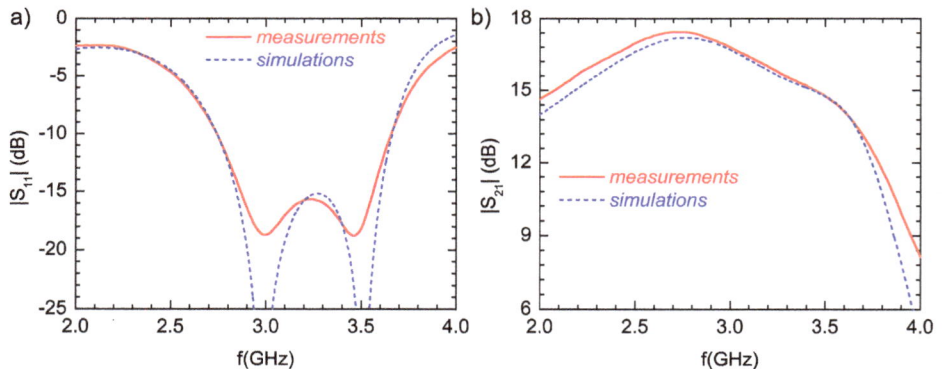

Figure 10. (**a**) the reflection coefficient $|S_{11}|$ and (**b**) the small-signal gain $|S_{21}|$ of the amplifier.

The power transfer characteristics $P_{out}(P_{in})$ of the amplifier measured for continuous wave (CW) mode from 2.5 GHz to 3.5 GHz frequency range are shown in Figure 11. Despite the gain irregularity, the available power is more equal. The output power at 1 dB gain compression (1 dB G.C.P.) point with associated gain vs. frequency is shown in Figure 12. In the whole frequency range, 1 dB G.C.P. is 30.3 ± 0.1 dBm.

The maximum output power of 32.2 dBm (1.66 W) was obtained at 3 GHz for the RF ON bias point V_{DS} = 28 V and I_{DS} = 150 mA (shallow class AB) of the transistor. Due to the high GaN HEMT chip thickness of 400 μm, the thermal conditions inside the transistor result in a drain current drop. It is the main limitation to obtaining higher output power level as well as higher efficiency. To reduce thermal resistance R_{th} value, the chip thickness should be decreased e.g., to standard used thickness of 100 μm. Additionally, source to ground connection is made using bond wires that decrease gain. Going through metal via holes from source pads to ground can improve the heat dissipation and gain, and reduce R_{on}.

The parameters of the amplifier show a high quality of fabricated transistors. The output power level is higher than 30 dB m at 1 dB G.C.P. over a 2.5 GHz to 3.5 GHz frequency range with 13.5 ± 1.0 dB gain. The maximum output power density of 4.15 W/mm is comparable with commercially available GaN HEMTs e.g., Wolfspeed CGH6008D 3.8 W/mm (Durham, NC, USA) [40], Qorvo TGF2023-2-01 4.8 W/mm) (Greensboro, NC, USA) [41].

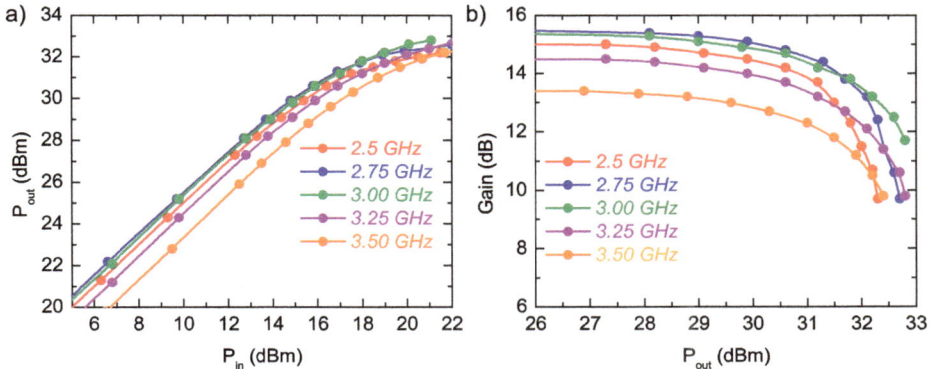

Figure 11. Transfer characteristics for continuous wave (CW) excitation—(**a**) output power $P_{out} = fP_{(in)}$ and (**b**) gain $G = f(P_{OUT})$.

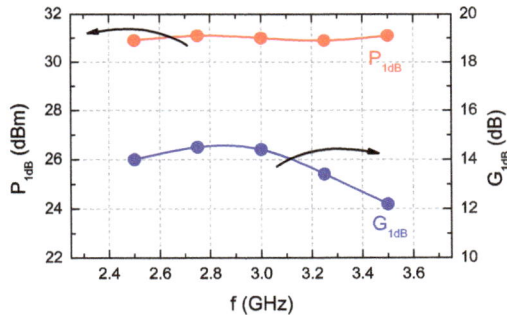

Figure 12. The output power at 1-dB gain compression point with associated gain of the amplifier.

4. Conclusions

In this work, AlGaN/GaN HEMTs, with active layers homoepitaxially grown on semi-insulating Ammono-GaN substrates, were fabricated. The use of regrown, highly-doped $In_xGa_{1-x}N/GaN$ sub-contact layers resulted in decreasing the contact resistivity from 0.8–1.1 $\Omega \cdot$mm (as for recessed ohmic contacts) to 0.3–0.6 $\Omega \cdot$mm and decreasing value of parasitic elements like source and drain resistance (see Table 1). This leads to enhancement of DC and RF performance as compared to AlGaN/GaN HEMTs on semi-insulating Ammono-GaN substrates with recessed Ti/Al/Mo/Au ohmic contacts [23]. An 1 A/mm on-state current was achieved and f_T and f_{MAX} were 21 and 30 GHz, respectively, for 0.8 µm gate length devices. Moreover, thanks to the use of selectively regrown highly-doped $In_xGa_{1-x}N/GaN$:Si layers, it is possible to use new types of metallization, e.g., with potentially increased thermal stability like Ru-based metallization schemes or to reduce the thermal budget to obtain low resistivity ohmic contacts, which is not possible to achieve using conventional technologies. Those ohmic contacts exhibit very good electrical parameters (R_c = 0.38–0.43 $\Omega \cdot$mm) and allow for obtaining transistors with DC and high-frequency parameters comparable to those for devices with Ti/Al/Mo/Au ohmic contacts with subcontact n^+-$In_xGa_{1-x}N$:Si regrown epilayer, annealed at high temperature of 850 °C.

Table 1. Comparison of ohmic contact resistances (R_C), and source and drain resistances (R_S and R_D) obtained from S-parameters of AlGaN/GaN high electron mobility transistor (HEMT) on SI Ammono-GaN substrates with Ti/Al/Mo/Au ohmic contacts with subcontact n^+-In$_x$Ga$_{1-x}$N:Si regrown epilayer and with recessed Ti/Al/Mo/Au ohmic contact.

Ohmic Contact	R_C ($\Omega \cdot$mm)	R_S (Ω)	R_D (Ω)
recessed Ti/Al/Mo/Au	0.8–1.1	0.86	3.4
Ti/Al/Mo/Au with regrown epilayer	0.3–0.6	0.31	1.78

Overall, we have shown AlGaN/GaN high electron mobility transistors fabricated on truly bulk monocrystalline semi-insulating GaN substrates with parameters (output current and output power density) comparing well with devices manufactured on silicon carbide substrates available from commercial manufacturers. The high potential of developed technology was proved by assembly of microwave systems using our devices (S-band power amplifiers). The fabricated S-band power amplifiers with AlGaN/GaN HEMTs on semi-insulating Ammono-GaN substrates have maximum power density of 4.15 W/mm. The high power and high frequency performance can be further enhanced by optimizing gate design and length and formation of via hole interconnections.

Author Contributions: W.W., M.G., and D.G. designed, fabricated and performed characterization of the microwave amplifier as well as measurements and modelling of HEMTs; M.Z. and R.K., performed ammonothermal growth of GaN substrates; P.P. performed epitaxial growth of AlGaN/GaN HEMT structures and regrowth of n-GaN for ohmic contacts; A.P., M.E., and E.K. conceived, designed and implemented building blocks and process flow of AlGaN/GaN HEMT processing technology, A.T. performed computer simulations and DC characterization of HEMT devices, and M.W. performed structural characterizations.

Funding: The research was supported by the National Centre for Research and Development PolHEMT Project, Contract No. PBS1/A3/9/2012.

Conflicts of Interest: The authors declare no conflict of interest.

References

1. Roccaforte, F.; Fiorenza, P.; Greco, G.; Lo Nigro, R.; Giannazzo, F.; Iucolano, F.; Saggio, M. Emerging trends in wide band gap semiconductors (SiC and GaN) technology for power devices. *Microelectron. Eng.* **2018**, *187–188*, 66–77. [CrossRef]

2. Pengelly, R.S.; Wood, S.M.; Milligan, J.W.; Sheppard, S.T.; Pribble, W.L. A Review of GaN on SiC High Electron-Mobility Power Transistors and MMICs. *IEEE Trans. Microw. Theory Tech.* **2012**, *60*, 1764–1783. [CrossRef]

3. Amano, H. Progress and Prospect of the Growth of Wide-Band-Gap Group III Nitrides: Development of the Growth Method for Single-Crystal Bulk GaN. *Jpn. J. Appl. Phys.* **2013**, *52*, 050001. [CrossRef]

4. Bockowski, M.; Iwinska, M.; Amilusik, M.; Fijalkowski, M.; Lucznik, B.; Sochacki, T. Challenges and future perspectives in HVPE-GaN growth on ammonothermal GaN seeds. *Semicond. Sci. Technol.* **2016**, *31*, 093002. [CrossRef]

5. Sochacki, T.; Bryan, Z.; Amilusik, M.; Bobea, M.; Fijalkowski, M.; Bryan, I.; Lucznik, B.; Collazo, R.; Weyher, J.L.; Kucharski, R.; et al. HVPE-GaN grown on MOCVD-GaN/sapphire template and ammonothermal GaN seeds: Comparison of structural, optical, and electrical properties. *J. Cryst. Growth* **2014**, *394*, 55–60. [CrossRef]

6. Sochacki, T.; Bryan, Z.; Amilusik, M.; Collazo, R.; Lucznik, B.; Weyher, J.L.; Nowak, G.; Sadovyi, B.; Kamler, G.; Kucharski, R.; et al. Preparation of Free-Standing GaN Substrates from Thick GaN Layers Crystallized by Hydride Vapor Phase Epitaxy on Ammonothermally Grown GaN Seeds. *Appl. Phys. Express* **2013**, *6*, 075504. [CrossRef]

7. Zajac, M.; Kucharski, R.; Grabianska, K.; Gwardys-Bak, A.; Puchalski, A.; Wasik, D.; Litwin-Staszewska, E.; Piotrzkowski, R.; Domagala, J.Z.; Bockowski, M. Basic ammonothermal growth of Gallium Nitride—State of the art, challenges, perspectives. *Prog. Cryst. Growth Charact. Mater.* **2018**, *64*, 63–74. [CrossRef]

8.   Dwiliński, R.; Doradziński, R.; Garczyński, J.; Sierzputowski, L.; Puchalski, A.; Kanbara, Y.; Yagi, K.; Minakuchi, H.; Hayashi, H. Excellent crystallinity of truly bulk ammonothermal GaN. *J. Cryst. Growth* **2008**, *310*, 3911–3916. [CrossRef]

9.   Cheney, D.J.; Douglas, E.A.; Liu, L.; Lo, C.F.; Xi, Y.Y.; Gila, B.P.; Ren, F.; Horton, D.; Law, M.E.; Smith, D.J.; Pearton, S.J. Reliability studies of AlGaN/GaN high electron mobility transistors. *Semicond. Sci. Technol.* **2013**, *28*, 074019. [CrossRef]

10.   Kuball, M.; Tapajna, M.; Simms, R.J.; Faqir, M.; Mishra, U.K. AlGaN/GaN HEMT device reliability and degradation evolution: Importance of diffusion processes. *Microelectron. Reliab.* **2011**, *51*, 195–200. [CrossRef]

11.   Moroz, V.; Wong, H.Y.; Choi, M.; Braga, N.; Mickevicius, R.V.; Zhang, Y.; Palacios, T. The Impact of Defects on GaN Device Behavior: Modeling Dislocations, Traps, and Pits. *ECS J. Solid State Sci. Technol.* **2016**, *5*, P3142–P3148. [CrossRef]

12.   Pearton, S.J.; Ren, F.; Patrick, E.; Law, M.E.; Polyakov, A.Y. Review—Ionizing Radiation Damage Effects on GaN Devices. *ECS J. Solid State Sci. Technol.* **2016**, *5*, Q35–Q60. [CrossRef]

13.   Sarua, A.; Ji, H.; Hilton, K.P.; Wallis, D.J.; Uren, M.J.; Martin, T.; Kuball, M. Thermal Boundary Resistance Between GaN and Substrate in AlGaN/GaN Electronic Devices. *IEEE Trans. Electron Devices* **2007**, *54*, 3152–3158. [CrossRef]

14.   Kuzmík, J.; Bychikhin, S.; Pogany, D.; Gaquière, C.; Pichonat, E.; Morvan, E. Investigation of the thermal boundary resistance at the III-Nitride/substrate interface using optical methods. *J. Appl. Phys.* **2007**, *101*, 054508. [CrossRef]

15.   Killat, N.; Montes, M.; Pomeroy, J.W.; Paskova, T.; Evans, K.R.; Leach, J.; Li, X.; Ozgur, U.; Morkoc, H.; Chabak, K.D.; et al. Thermal Properties of AlGaN/GaN HFETs on Bulk GaN Substrates. *IEEE Electron Device Lett.* **2012**, *33*, 366–368. [CrossRef]

16.   Kruszewski, P.; Prystawko, P.; Kasalynas, I.; Nowakowska-Siwinska, A.; Krysko, M.; Plesiewicz, J.; Smalc-Koziorowska, J.; Dwilinski, R.; Zajac, M.; Kucharski, R.; et al. AlGaN/GaN HEMT structures on ammono bulk GaN substrate. *Semicond. Sci. Technol.* **2014**, *29*, 075004. [CrossRef]

17.   Anderson, T.J.; Tadjer, M.J.; Hite, J.K.; Greenlee, J.D.; Koehler, A.D.; Hobart, K.D.; Kub, F.J. Effect of Reduced Extended Defect Density in MOCVD Grown AlGaN/GaN HEMTs on Native GaN Substrates. *IEEE Electron Device Lett.* **2016**, *37*, 28–30. [CrossRef]

18.   Krupka, J.; Judek, J. RF Capacitive Spectroscopy for Contactless Measurements of Resistivity Profiles in Highly Resistive Semiconductor Wafers. *IEEE Trans. Semicond. Manuf.* **2014**, *27*, 530–538. [CrossRef]

19.   Krupka, J.; Zając, M.; Kucharski, R.; Gryglewski, D. Dielectric properties of highly resistive GaN crystals grown by ammonothermal method at microwave frequencies. *AIP Adv.* **2016**, *6*, 035313. [CrossRef]

20.   Krupka, J. Contactless methods of conductivity and sheet resistance measurement for semiconductors, conductors and superconductors. *Meas. Sci. Technol.* **2013**, *24*, 062001. [CrossRef]

21.   Simon, R.B.; Anaya, J.; Kuball, M. Thermal conductivity of bulk GaN—Effects of oxygen, magnesium doping, and strain field compensation. *Appl. Phys. Lett.* **2014**, *105*, 202105. [CrossRef]

22.   Joglekar, S.; Azize, M.; Beeler, M.; Monroy, E.; Palacios, T. Impact of recess etching and surface treatments on ohmic contacts regrown by molecular-beam epitaxy for AlGaN/GaN high electron mobility transistors. *Appl. Phys. Lett.* **2016**, *109*, 041602. [CrossRef]

23.   Taube, A.; Kamińska, E.; Kozubal, M.; Kaczmarski, J.; Wojtasiak, W.; Jasiński, J.; Borysiewicz, M.A.; Ekielski, M.; Juchniewicz, M.; Grochowski, J.; et al. Ion implantation for isolation of AlGaN/GaN HEMTs using C or Al. *Phys. Status Solidi A* **2015**, *212*, 1162–1169. [CrossRef]

24.   Greco, G.; Iucolano, F.; Bongiorno, C.; Di Franco, S.; Lo Nigro, R.; Giannazzo, F.; Prystawko, P.; Kruszewski, P.; Krysko, M.; Grzanka, E.; et al. Electrical and structural properties of Ti/Al-based contacts on AlGaN/GaN heterostructures with different quality. *Phys. Status Solidi A* **2015**, *212*, 1091–1098. [CrossRef]

25.   Wang, L.; Mohammed, F.M.; Adesida, I. Differences in the reaction kinetics and contact formation mechanisms of annealed Ti/Al/Mo/Au Ohmic contacts on n-GaN and AlGaN/GaN epilayers. *J. Appl. Phys.* **2007**, *101*, 013702. [CrossRef]

26.   Wang, L.; Kim, D.H.; Adesida, I. Direct contact mechanism of Ohmic metallization to AlGaN/GaN heterostructures via Ohmic area recess etching. *Appl. Phys. Lett.* **2009**, *95*, 172107. [CrossRef]

27.   Wang, L.; Mohammed, F.M.; Adesida, I. Formation mechanism of Ohmic contacts on AlGaN/GaN heterostructure: Electrical and microstructural characterizations. *J. Appl. Phys.* **2008**, *103*, 093516. [CrossRef]

28. Schroder, D.K. Contact Resistance and Schottky Barriers. In *Semiconductor Material and Device Characterization*; John Wiley & Sons, Inc.: Hoboken, NJ, USA, 2005.

29. Guo, J.; Li, G.; Faria, F.; Cao, Y.; Wang, R.; Verma, J.; Gao, X.; Guo, S.; Beam, E.; Ketterson, A.; et al. MBE-Regrown Ohmics in InAlN HEMTs With a Regrowth Interface Resistance of 0.05 Ω mm. *IEEE Electron Device Lett.* **2012**, *33*, 525–527. [CrossRef]

30. Kaczmarski, J.; Grochowski, J.; Kaminska, E.; Taube, A.; Dyczewski, J.; Jung, W.; Dynowska, E.; Piotrowska, A. Transparent Amorphous Ru–Si–O Schottky Contacts to In–Ga–Zn–O. *J. Disp. Technol.* **2015**, *11*, 528–532. [CrossRef]

31. Kaczmarski, J.; Borysiewicz, M.A.; Piskorski, K.; Wzorek, M.; Kozubal, M.; Kamińska, E. Flexible IGZO Schottky diodes on paper. *Semicond. Sci. Technol.* **2018**, *33*, 015010. [CrossRef]

32. Kaminska, E.; Piotrowska, A.; Szczesny, A.; Kuchuk, A.; Lukasiewicz, R.; Golaszewska, K.; Kruszka, R.; Barcz, A.; Jakiela, R.; Dynowska, E.; et al. Thermally stable Ru-Si-O gate electrode for AlGaN/GaN HEMT. *Phys. Status Solidi C* **2005**, *2*, 1060–1064. [CrossRef]

33. Brunel, L.; Malbert, N.; Curutchet, A.; Labat, N.; Lambert, B. Kink effect characterization in AlGaN/GaN HEMTs by DC and drain current transient measurements. In Proceedings of the European Solid-State Device Research Conference (ESSDERC), Bordeaux, France, 17–21 September 2012; pp. 270–273.

34. Nigam, A.; Bhat, T.N.; Rajamani, S.; Dolmanan, S.B.; Tripathy, S.; Kumar, M. Effect of self-heating on electrical characteristics of AlGaN/ GaN HEMT on Si (111) substrate. *AIP Adv.* **2017**, *7*, 085015. [CrossRef]

35. Gaska, R.; Osinsky, A.; Yang, J.; Shur, M. Self-heating in high-power AlGaN-GaN HFETs. *IEEE Electron Device Lett.* **1998**, *19*, 89–91. [CrossRef]

36. Gummel, H.K. On the definition of the cutoff frequency f_T. *Proc. IEEE* **1969**, *57*, 2159–2159. [CrossRef]

37. Walker, J. *High-Power GaAs FET Amplifiers*; Artech House, Inc.: Norwood, MA, USA, 1993.

38. Wojtasiak, W.; Gryglewski, D. A 100 W SiC MESFET Amplifier for L-band T/R Module of APAR. *Int. J. Electron. Telecommun.* **2011**, *57*, 135–140. [CrossRef]

39. Gryglewski, D.; Wojtasiak, W.; Morawski, T. Phased arrays with distributed feed. In *Modern Radar Technology. Controllable Beam Antennas in Radar Technology*; Transport and Communication Publishers: Warsaw, Poland, 2012; pp. 189–224.

40. Available online: http://www.wolfspeed.com (accessed on 3 September 2018).

41. Available online: https://www.qorvo.com (accessed on 29 August 2018).

micromachines

MDPI

Article

Investigation on the I–V Kink Effect in Large Signal Modeling of AlGaN/GaN HEMTs

Shuman Mao and Yuehang Xu *

School of Electronic Science and Engineering (National Exemplary School of Microelectronics),
University of Electronic Science and Technology of China, Chengdu 611731, China;
201711020135@std.uestc.edu.cn
* Correspondence: yuehangxu@uestc.edu.cn

Received: 26 September 2018; Accepted: 2 November 2018; Published: 5 November 2018

check for
updates

Abstract: The effect brought by the I–V kink effect on large signal performance of AlGaN/GaN high electron mobility transistors (HEMTs) was investigated in this paper. An improved compact model was proposed to accurately characterize the I–V kink effect. The bias dependence of the I–V kink effect has also been taken into consideration. AlGaN/GaN HEMTs with different gate width were utilized to validate the proposed model. Built on the proposed model, the effect brought by the I–V kink effect on large signal performance has been studied. Results show that the I–V kink effect will lead to the degradation of characteristics, including output power, gain, and power-added efficiency at the saturation region. Furthermore, the influence of the I–V kink effect was found to be related with the input power and the static bias point in this work. The time domain waveform and AC dynamic load line were used for validation of results based on simulation. The consequences of this paper will be useful for the optimization of practical circuit design.

Keywords: I–V kink effect; AlGaN/GaN HEMT; large signal performance

1. Introduction

Due to the special characteristics of the material itself, gallium nitride (GaN) has been widely used in wireless applications, THz band emerging devices, space industry, power electronics, and many other fields [1–3]. Also, with the development of fabrication techniques, the feature size of GaN-based devices have been shrinking to less than 100 nm [4]. Together with its unique characteristics, especially higher breakdown voltage and higher output power density, it has been proven to be an excellent candidate in high frequency applications [5]. Numerous circuit designs based on GaN processes with outstanding performance have sprung out these recent years [6–8]. The rapid development of GaN-based devices also stimulates the improvement of compact modeling, which serves as a key to practical circuit design [9].

In the past few years, lots of work has been focused on the characterization of electrothermal [10–12] and trapping effects [13–15] in compact modeling of GaN high electron mobility transistors (HEMTs). Many compact models, such as EE_HEMT1 model [16], Angelov GaN model [17], MVSG model [18], and ASM GaN model [19], have been developed for accurate characterization of device performance. Besides the development of the core model mentioned above, other real device effect models, such as noise model [20] and gate current model [19], are also proposed to improve the core model. I–V kink effect is also a common phenomenon observable in several kinds of transistors [21]. The mechanism has been thoroughly studied in many works. The setting of maximum value of drain-source voltage [22], as well as the sweeping direction of drain-source voltage in measurement [23], will also affect the kink degree. Also, the influence brought by the I–V kink effect on S parameters has been studied in [24]. However, the characterization of the I–V kink effect in compact model [25], and the influence brought by it on the large signal performance, have seldom been reported. As the I–V kink effect is observable when the device is biased at linear region, the load line will be influenced by it. In terms of practical circuit design, especially for high linearity and many other applications, it is worthy to develop an accurate I–V kink effect model and investigate the influence brought by it on large signal performance.

In this paper, the effect brought by the I–V kink effect on large signal performance of AlGaN/GaN HEMTs was studied. An improved compact model was proposed to accurately characterize the I–V kink effect. The model was validated via GaN HEMTs with different gate widths. Built on the proposed model, the effect brought by the I–V kink effect on large signal performance has been studied with different input power and under different bias points, separately. The time domain waveform and AC dynamic load line were also used for validation of results based on simulation in this work.

This paper is organized as follows. In Section 2, the model description was presented. The modeling method of the I–V kink effect was given in detail. In Section 3, the proposed model was validated by two AlGaN/GaN HEMTs with different gate width, at first. Then, the influence brought by the I–V kink effect with different input power and under different bias points was analyzed separately. Finally, in Section 4, the conclusion of this work is presented.

2. Model Description

2.1. Characterization of the I–V Kink Effect

The I–V kink effect is a common physical phenomenon in several kinds of transistors. Trapping effects have been proved to be the main reason for the occurrence of I–V kink effect [26]. Due to the defect induced by fabrication, the trap distributed in devices will lead to the current collapse when devices are biased at linear region [27]. However, along with the increase of drain-source voltage, the rise of the electric field in the channel will assist the de-trapping process. This will, in the end, lead to the "jump" of drain-source current [28]. The drain-source voltage V_{ds}, when the current recovers, is called V_{ds_kink} in numerous works. This phenomenon can also be captured in the AlGaN/GaN HEMTs used in this work. I–V curves of GaN HEMTs with different gate width are presented in Figure 1. It is worthy to mention here that the AlGaN/GaN HEMTs used in this work were fabricated in WIN SEMICONDUCTORS Corporation NP25-00 Gallium Nitride process. These devices were grown on a 4 mil and 100 µm thickness SiC substrate.

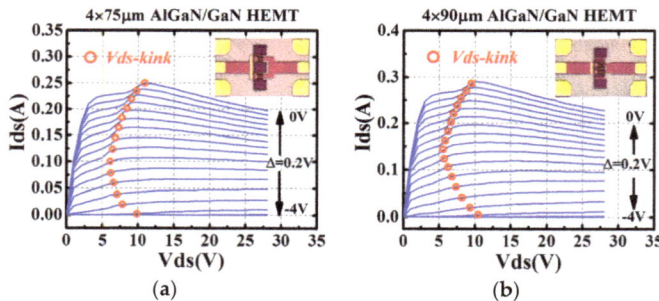

Figure 1. I–V kink effect in 0.25 µm AlGaN/GaN HEMTs with different gate width: (**a**) 4 × 75 µm and (**b**) 4 × 90 µm.

As the I–V kink effect only leads to the current collapse when V_{ds} is lower than 10 V in Figure 1, modification is needed to revise the nonlinear current formulation in the conventional compact model. In order to improve the convergence of the compact model, a simplified version based on the method in [25] was employed, in this work, to characterize the kink effect. The parameter V_{dsi} in [25] was not found to be suitable in this work, and was replaced by drain-source voltage V_{ds} to simplify the model.

Also, as the parameter V_{ds_k0} in [25] directly determines the value of V_{ds_kink} in Figure 1, it has been changed to V_{ds_kink} in this work. The formulation of the whole I–V kink model is shown in Equation (1).

$$I_{kink} = I_k \times \left(1 + \tanh\left(\frac{V_{ds} - V_{ds\_kink}}{V_{ds\_k1}}\right)\right), \tag{1a}$$

$$I_k = I_{k0} \times \exp\left(-\left(\frac{V_{gs} - V_{gs\_k0}}{V_{gs\_k1}}\right)^2\right), \tag{1b}$$

where I_{k0}, V_{ds_k1}, V_{ds_kink}, V_{gs_k0}, and V_{gs_k1} are parameters which can be achieved by measured I–V curve.

In Equation 1, I_{k0} denotes the maximum difference between the I_{ds} curve with I–V kink effect, and the one without I–V kink effect [25]. The V_{gs_k0} refers to the gate-source voltage V_{gs} when I_k is equal to I_{k0}. Since the GaN HEMT in this work is a kind of depletion mode device which is mainly for power amplifier application, gate-source voltage V_{gs} is equal to or less than zero. As a result, V_{gs_k0} is equal to zero in this work. V_{gs_k1} is a parameter used for describing the variation of I_k under different V_{gs}. It can be achieved by fitting I_k under different V_{gs}. V_{ds_k1} denotes the slope of the I–V curve transferring from the kink region to the region without kink. V_{ds_kink} refers to the drain-source voltage when the I–V kink effect disappears in the I–V curve.

However, with the model in Equation (1), the value of V_{ds_kink} remains the same under each V_{gs}. The bias dependence of V_{ds_kink} cannot be accurately characterized only with Equation (1) because the parameter V_{ds_kink} is not an expression related to the gate-source voltage V_{gs}. As the variation of V_{ds_kink}, along with the change of V_{gs} shown in Figure 1, agrees well with the trend of a cubic function, the formulation in (2) was used in this work to add bias dependence into parameter V_{ds_kink} in Equation (1).

$$V_{ds\_kink} = aV_{gs}^3 + bV_{gs}^2 + cV_{gs} + d, \tag{2}$$

where a, b, c, and d are all fitting parameters. They can be achieved by polynomial fitting with the extracted discrete V_{ds_kink} under different V_{gs}, which have been marked in red circles in Figure 1.

With the parameter extraction method mentioned above, the I–V kink effect model can be achieved. The extracted parameters of the I–V kink effect model in Equations (1) and (2) for the 4×75 μm AlGaN/GaN HEMT in Figure 1a are presented in Table 1.

Table 1. The extracted I–V kink effect model parameters in Equations (1) and (2).

I_{k0}	v_{ds_k1}	a	b	v_{gs_k0}	v_{gs_k1}	c	d
0.04	0.012	−0.7	−1.01	0	3.08	2.51	9.82

2.2. Compact Modeling and Its Validation

Comparing with other physical based [18,19] or empirical [16] compact model, the Angelov model [17] takes advantage of good convergence and much easier parameter extraction. As a result, an empirical compact modeling method [29] based on the Angelov theory was used in this work to model the drain-source current of the GaN HEMT. With the modification based on Equations (1) and (2), the bias dependence of the I–V kink effect can be accurately described. The self-heating effect was modeled by the variation of channel temperature and the trapping effect was modeled by the equivalent gate voltage method [15] in the compact model of this work. Then, the I–V kink effect model is integrated into the compact model by direct addition into the current model expression shown in Equation (3a). The proposed model in this work is also scalable. The capacitance model, including C_{gs} and C_{gd} mentioned in [29], is used in this work.

$$I_{ds} = I_{ds\_nkink} + I_{kink}, \tag{3a}$$

$$I_{ds\_nkink} = I_{pkth} \times \left(1 + M_{ipkth} \times \tanh(\psi)\right) \times \tanh(\alpha V_{ds}), \tag{3b}$$

where I_{pkth}, M_{ipkth}, ψ, and α are all model parameters of the improved Angelov model in [29]. I_{kink} denotes the expression in Equation (1a).

The static DCIV curves in this work were measured on cascade deck (Summit 12000, FormFactor, Livermore, CA, USA) with the help of power device analyzer/curve tracer (Keysight B1505A, Keysight Technologies, Santa Rosa, CA, USA). The photography of the on-wafer measurement system is shown in Figure 2a.

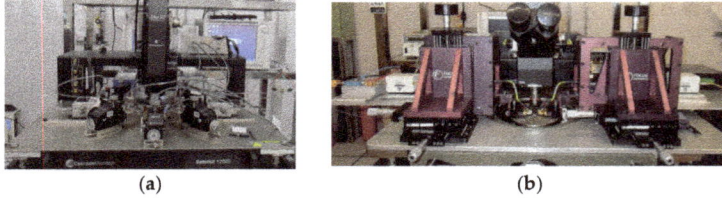

(a)   (b)

Figure 2. The on-wafer measurement system. (**a**) The measurement desk for DCIV and S parameter; (**b**) The on-wafer load-pull system.

The on-wafer load-pull system used for achieving maximum output power is shown in Figure 2b. The measurement is performed on Cascade Summit 11000 (FormFactor, Livermore, CA, USA). The testing block diagram is the same as the one mentioned in [30]. The main instruments in this work include the source and load tuner (Focus CCMT-5080, Focus Microwaves Inc., Québec City, QC, Canada), vector network analyzer (Agilent N5245A, Keysight Technologies, Santa Rosa, CA, USA), DC power (Agilent E3633A/E3634A, Keysight Technologies, Santa Rosa, CA, USA), and the input signal amplifier (Agilent 83020A, Keysight Technologies, Santa Rosa, CA, USA). S parameters in this work are measured with Agilent N5247 (Keysight Technologies, Santa Rosa, CA, USA), and the DCIV curves are measured with Keysight B1505A (Keysight Technologies, Santa Rosa, CA, USA).

In order to validate the proposed model, two AlGaN/GaN HEMTs with the gate width of 4×75 μm and 4×90 μm were used. The flow chart for validation procedure is presented in Figure 3.

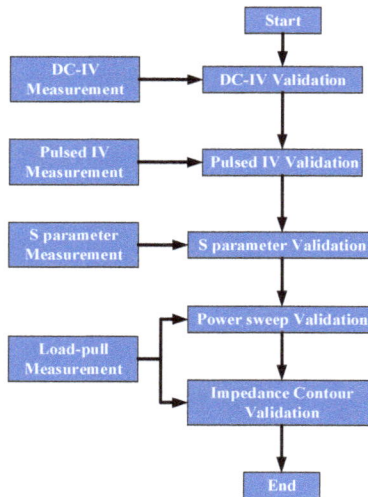

Figure 3. Flow chart of the validation procedure of the proposed model.

The comparison between simulation results and measured data of the DCIV of 4 × 75 μm and 4 × 90 μm GaN HEMTs at room temperature are shown in Figure 4. The gate-source voltage V_{gs} is swept from −4 V to 0 V stepped by 0.2 V, and drain-source voltage V_{ds} is from 0 V to 28 V, stepped by 1 V to include the I–V kink effect.

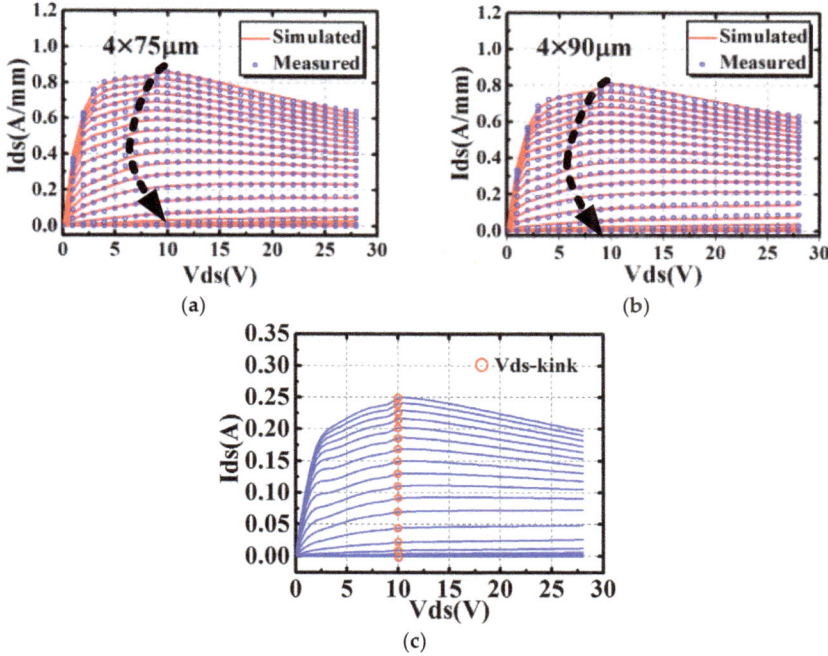

Figure 4. Validation of DCIV characteristics of the proposed model: (a) 4 × 75 μm GaN HEMT, (b) 4 × 90 μm GaN HEMT and (c) Results of the modeling approach in [25].

Comparing with Figure 4c, it is clear that the bias dependence of $V_{ds\text{-}kink}$ in Figure 4a,b can be accurately described under different gate-source voltage V_{gs} based on the proposed bias dependence model in Equation (2).

3. Investigation on Large Signal Performance

3.1. Validation of the Large Signal Model

The large signal model was embedded into Keysight ADS by symbolically defined device (SDD) tool. The small signal characteristics of the model have been validated at first. Results for S parameters at V_{gs} = −2.2 V, V_{ds} = 20 V and V_{gs} = −2.6 V, V_{ds} = 28 V are presented in Figure 5. The frequency band is from 0.1 GHz to 40 GHz in Figure 5.

It can be observed in Figure 5 that the proposed model can accurately characterize the small signal characteristics under different bias points over the frequency band. Then, in order to validate the large signal characteristics of the model, on-wafer load-pull measurement was performed to achieve the impedance of maximum output power. The impedance contours of a 4 × 75 μm GaN HEMT for maximum output power and maximum power-added efficiency (PAE) are presented in Figure 6.

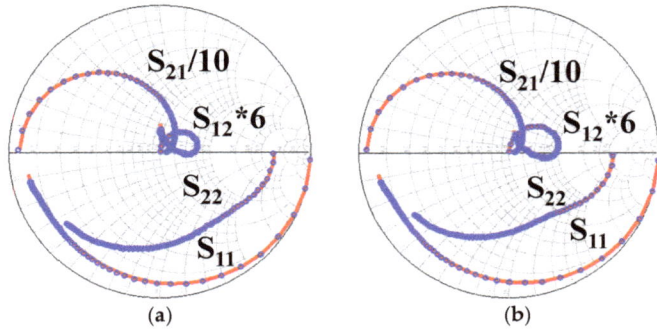

Figure 5. Small signal characteristic validation of the proposed model at different bias points: (a) $V_{gs} = -2.2$ V, $V_{ds} = 20$ V and (b) $V_{gs} = -2.6$ V, $V_{ds} = 28$ V.

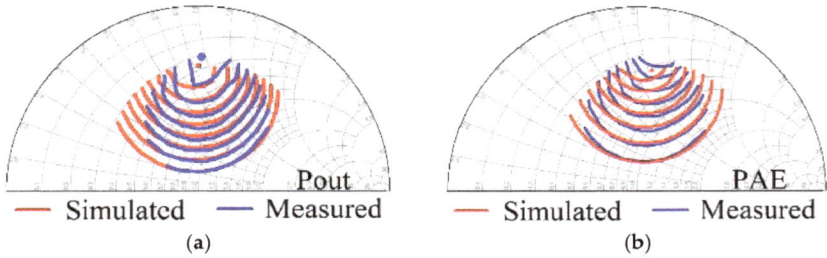

Figure 6. Validation of the impedance contours of the proposed model: (a) Impedance contour for maximum output power and (b) Impedance contour for maximum PAE.

It is clear in Figure 6 that the proposed model can accurately predict the impedance contours. Terminated at the optimum source and load impedance, power sweep characteristics at a certain bias and frequency point can be achieved. The power sweep characteristics, including output power (Pout), gain, and power-added efficiency (PAE), are also validated for the AlGaN/GaN HEMTs with gate width of 4×75 μm and 4×90 μm. Results when the frequency of the input signal is 10 GHz are presented in Figure 7.

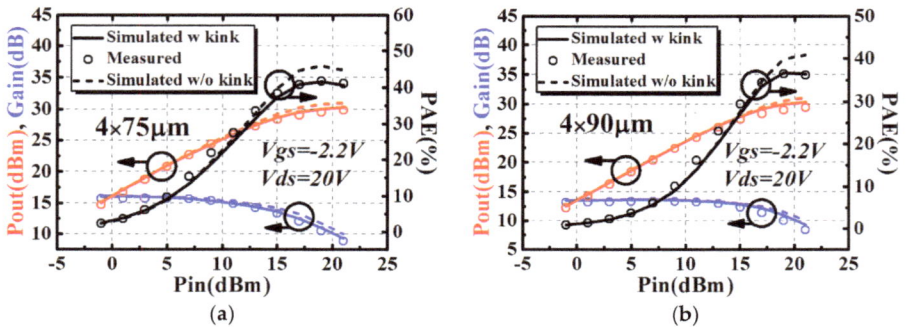

Figure 7. The influence brought by the I–V kink effect on large signal performance of AlGaN/GaN HEMTs @10GHz: (a) 4×75 μm and (b) 4×90 μm.

The influence brought by the I–V kink effect on large signal performance was also studied in Figure 7. The comparison between two conditions when one has taken the I–V kink effect into consideration, while the other has not, are shown in Figure 7. It can be seen in Figure 7 that the I–V kink effect will lead to the degradation of large signal performance, including output power (Pout), gain, and power-added efficiency (PAE) at the saturation region. In terms of output power and gain, the degradation induced by the I–V kink effect is only 1 dB or even smaller. This degradation is in an acceptable range for circuit design. However, the degradation of power-added efficiency is observable, comparing with output power and gain. This should be taken into consideration in practical circuit design, especially high efficiency applications.

Besides the large signal performance at a certain frequency point, the influence brought by the I–V kink effect on time domain characteristics was then studied by simulation. The 4×75 µm AlGaN/GaN HEMT was used for investigation. It was terminated in its optimum impedance for maximum output power at 10 GHz. The input impedance Z_s is $14.05 + j * 18.50$ Ω, while the output impedance Z_L is $30.37 + j * 48.27$ Ω. The input power was set to 14.5 dBm. Results are shown in Figure 8.

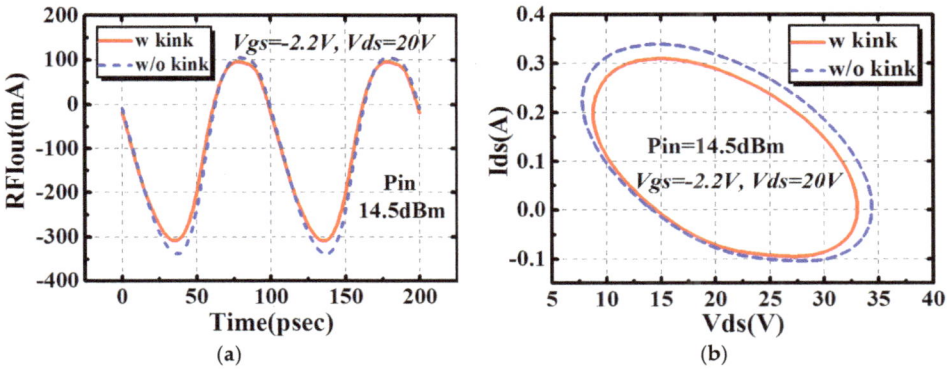

Figure 8. The influence brought by the I–V kink effect on time domain characteristics of AlGaN/GaN HEMTs @10GHz: (**a**) time domain waveform and (**b**) AC dynamic load line.

It is clear in Figure 8 that the I–V kink effect affects both the time domain waveform and the AC dynamic load line. Also, the I–V kink effect only affects the magnitude of output waveform. The phase remains the same in two conditions.

3.2. The Influence of I–V Kink Effect with Different Input Power

In order to further validate the influence brought by the I–V kink effect on large signal performance shown in Figure 7 on the aspect of input power, the time domain characteristics of the 4×75 µm AlGaN/GaN HEMT used in this work were studied. Characteristics, including the AC dynamic load line and output time domain waveform of the device with different input power, are shown in Figure 9. The input power was swept from 13 dBm to 21 dBm for investigation of different working states of the 4×75 µm GaN HEMT.

It can be observed in Figure 9a,b that, with the increase of input power, the influence brought by the I–V kink effect aggravates for both time domain waveform and AC dynamic load line. The decrease of the amplitude of the waveform will, in the end, lead to the degradation of output performance presented in Figure 7. The decrease of the amplitude of the waveform can be further explained by the distribution of AC dynamic load lines with different input power, shown in Figure 9c. Along with the rise of input power, more parts of the trace of AC dynamic load line will be located in the kink region marked in the dashed box in Figure 9c. Nonlinear effects will be induced by the current collapse in the region at this time. As a result, the I–V kink effect-induced current collapse should be considered the main reason for the variation shown in Figure 9a,b. Based on the analysis above, a compromising input power should be chosen in circuit design if the I–V kink effect is observable in the transistors.

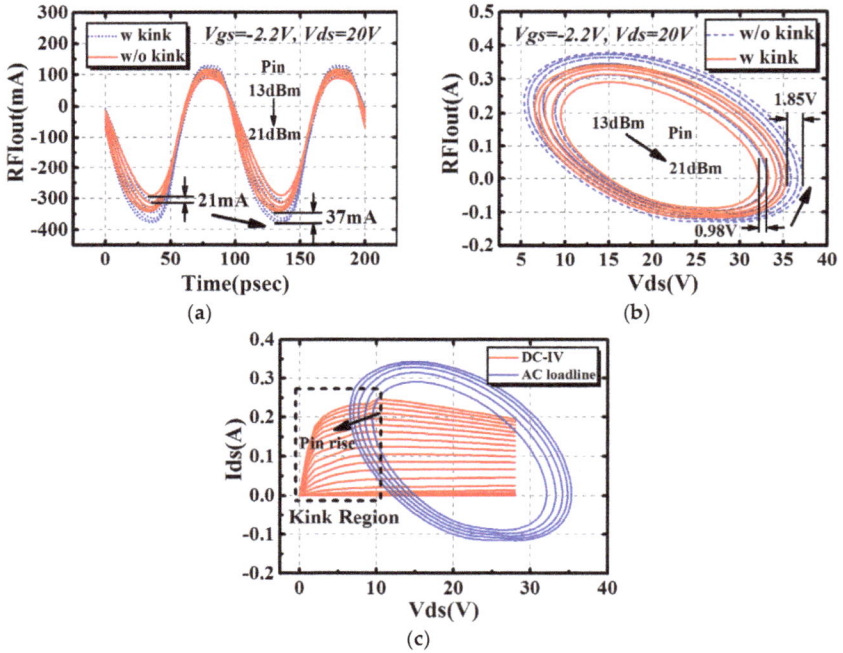

Figure 9. Investigation on influence brought by the I–V kink effect on time domain characteristics with different input power @10 GHz: (**a**) time domain waveform, (**b**) AC dynamic load line and (**c**) DCIV and AC dynamic load line.

3.3. The Influence of I–V Kink Effect under Different Bias Points

As the static bias point of a certain device also affects the distribution of AC dynamic load line, the bias dependence of the influence brought by the I–V kink effect has also been studied. The large signal performance at bias points, including $V_{gs} = -2.6$ V, $V_{ds} = 20$ V, $V_{gs} = -2.6$ V, $V_{ds} = 28$ V, $V_{gs} = -2.2$ V, and $V_{ds} = 28$ V were investigated in Figure 10. The time domain waveforms for each bias point are also included. The input power chosen for investigation is 21 dBm because the device is saturated. The influence brought by the I–V kink effect is observable under this circumstance.

The influence brought by the I–V kink effect on the amplitude of the waveform weakens along with the variation of bias points from Figure 10a–c. The variation trend in Figure 10 is the same as the one in Figure 7, that there is degradation of characteristics, including output power, gain, and power-added efficiency at the saturation region. Also, the degradation of PAE is observable compared with output power and gain. As the influence of the I–V kink effect on power-added efficiency (PAE) is observable, the influence has been further calculated by $\Delta PAE = PAEnk - PAEk$. Where $PAEnk$ refers to PAE calculated without the I–V kink effect, while $PAEk$ refers to PAE calculated with the I–V kink effect. Results for four different bias points are listed in Table 2.

Table 2 shows that the influence of the I–V kink effect on PAE is related with both gate bias and drain bias. This phenomenon can be further validated by the distribution of AC dynamic load lines at these bias points. The input power was also 21 dBm, which was the same as the one in Figure 10. Results are shown in Figure 11. The index 1, 2, 3, and 4 in Figure 11 refer to the four bias points, which are consistent with the ones in Table 2.

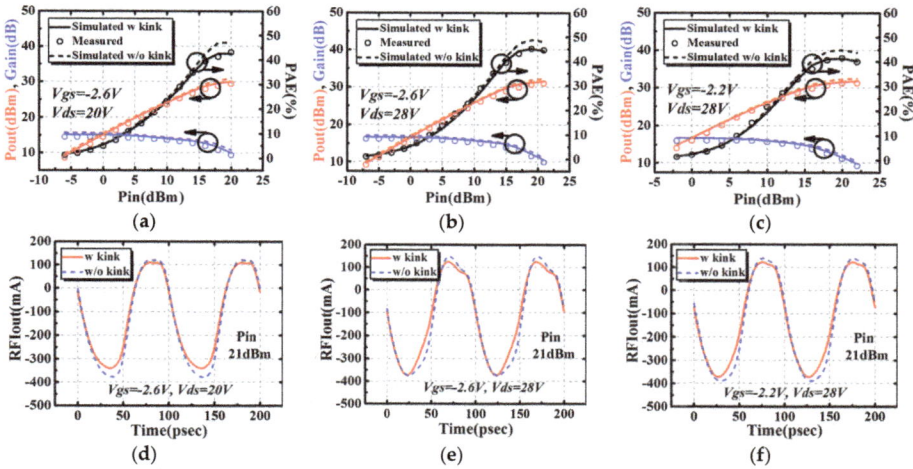

Figure 10. Investigation on the influence brought by the I–V kink effect at different bias points: (a–c) refers to the large signal output performance while (d–f) refers to the time domain waveform.

Table 2. The influence of I–V kink effect on power-added efficiency.

Index	Bias Point	ΔPAE
1	$V_{gs} = -2.2$ V, $V_{ds} = 28$ V	3.3%
2	$V_{gs} = -2.6$ V, $V_{ds} = 28$ V	3.8%
3	$V_{gs} = -2.2$ V, $V_{ds} = 20$ V	4.3%
4	$V_{gs} = -2.6$ V, $V_{ds} = 20$ V	5.1%

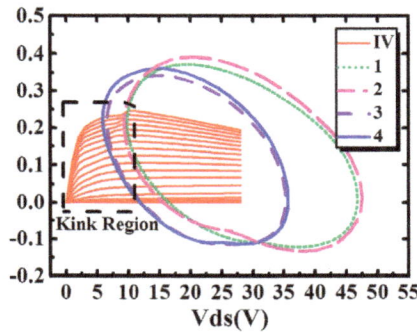

Figure 11. The distribution of AC dynamic load lines at four different bias points.

It can be observed in Figure 11 that, with the rise of index presented in Table 2, more parts of each trace will be located in the kink region marked in the dashed box. As a result, the influence brought by the I–V kink effect will aggravate, at the same time. It can be concluded that the bias dependence of the influence is directly dependent on the length of the trace located in the kink region.

4. Conclusions

In this paper, the effect brought by the I–V kink effect on large signal performance of AlGaN/GaN HEMTs was investigated. An improved compact model was proposed to accurately characterize the I–V kink effect. The effect brought by the I–V kink effect on large signal performance has been studied with different input powers and under different bias points. Results show that the I–V kink effect will lead to the degradation of characteristics, including output power, gain, and power-added efficiency at saturation region. The degradation of output power and gain is in an acceptable range, while the degradation of PAE should be taken into consideration in circuit design. Results of this paper will be useful for optimization of practical circuit design.

Author Contributions: Writing—S.M.; Methodology—Y.X.

Funding: This work is supported in part by National Natural Science Foundation of China (Grant No. 61474020), the Fundamental Research Funds for the Central Universities (Grant No. ZYGX2016J036), Special Zone of National Defense Science and Technology Innovation (Grant No. 1716313-ZD00100103), and National Science and Technology Major Project of China (Grant No. 2017ZX01001101).

Conflicts of Interest: The authors declare no conflicts of interest

References

1. Mishra, U.K.; Shen, L.; Kazior, T.E.; Wu, Y.F. GaN-based RF power devices and amplifiers. *Proc. IEEE* **2008**, *96*, 287–305. [CrossRef]
2. Ohno, Y.; Kuzuhara, M. Application of GaN-based heterojunction FETs for advanced wireless communication. *IEEE Trans. Electron Devices* **2001**, *48*, 517–523. [CrossRef]
3. Sokolov, V.N.; Kim, K.W.; Kochelap, V.A.; Woolard, D.L. Terahertz generation in submicron GaN diodes within the limited space-charge accumulation regime. *J. Appl. Phys.* **2005**, *98*, 3096. [CrossRef]
4. Wu, S.; Guo, J.; Wang, W.; Zhang, J. W-band MMIC PA with ultrahigh power density in 100-nm ALGaN/GaN technology. *IEEE Trans. Electron Devices* **2016**, *63*, 3882–3886. [CrossRef]
5. Wienecke, S.; Romanczyk, B.; Guidry, M.; Li, H.; Ahmadi, E.; Hestroffer, K.; Zheng, X.; Keller, S.; Mishra, U.K. N-polar GaN cap MISHEMT with record power density exceeding 6.5 W/mm at 94 GHz. *IEEE Electron. Device Lett.* **2017**, *38*, 359–362. [CrossRef]
6. Chen, K.; Peroulis, D. Design of highly efficient broadband class-E power amplifier using synthesized low-pass matching networks. *IEEE Trans. Microw. Theory Tech.* **2011**, *59*, 3162–3173. [CrossRef]
7. Kim, J.; Park, H.; Lee, S.; Kwon, Y. 6–18 GHz, 8.1 W size-efficient GaN distributed amplifier MMIC. *Electron. Lett.* **2016**, *52*, 622–624. [CrossRef]
8. Colantonio, P.; Giannini, F.; Giofre, R.; Piazzon, L. High-efficiency ultra-wideband power amplifier in GaN technology. *Electron. Lett.* **2008**, *44*, 130–131. [CrossRef]
9. Dunleavy, L.; Baylis, C.; Curtice, W.; Connick, R. Modeling GaN: Powerful but challenging. *IEEE Microw. Mag.* **2010**, *11*, 82–96. [CrossRef]
10. Jardel, O.; Groote, F.D.; Reveyrand, T.; Jacquet, J.; Charbonniaud, C.; Teyssier, J.; Floriot, D.; Quere, R. An electrothermal model for AlGaN/GaN power hemts including trapping effects to improve large-signal simulation results on high VSWR. *IEEE Trans. Microw. Theory Tech.* **2007**, *55*, 2660–2669. [CrossRef]
11. Liu, L.; Ma, J.; Ng, G. Electrothermal large-signal model of III–V FETs including frequency dispersion and charge conservation. *IEEE Trans. Microw. Theory Tech.* **2009**, *57*, 3106–3117. [CrossRef]
12. Wen, Z.; Xu, Y.; Chen, Y.; Tao, H.; Ren, C.; Lu, H.; Wang, Z.; Zheng, W.; Zhang, B.; Chen, T.; Gao, T.; Xu, R. A quasi-physical compact large-signal model for AlGaN/GaN HEMTs. *IEEE Trans. Microw. Theory Tech.* **2017**, *65*, 5113–5122. [CrossRef]
13. Binari, S.C.; Ikossi, K.; Roussos, J.A.; Kruppa, W.; Doewon, P.; Dietrich, H.B.; Koleske, D.D.; Wickenden, A.E.; Henry, R.L. Trapping effects and microwave power performance in algan/gan hemts. *IEEE Trans. Electron Devices* **2001**, *48*, 465–471. [CrossRef]
14. Binari, S.C.; Klein, P.B.; Kazior, T.E. Trapping effects in GaN and SiC microwave FETs. *Proc. IEEE* **2002**, *90*, 1048–1058. [CrossRef]
15. Yuk, K.S.; Branner, G.R.; McQuate, D.J. A wideband multiharmonic empirical large-signal model for high-power GaN HEMTs with self-heating and charge-trapping effects. *IEEE Trans. Microw. Theory Tech.* **2009**, *57*, 3322–3332. [CrossRef]
16. Lautensack, C.; Chalermwisutkul, S.; Jansen, R.H. Modification of EEHEMT1 Model for Accurate Description of GaN HEMT Output Characteristics. In Proceedings of the 2007 Asia-Pacific Microwave Conference, Bangkok, Thailand, 11–14 December 2007.

17. Angelov, I.; Thorsell, M.; Andersson, K.; Rorsman, N.; Kuwata, E.; Ohtsuka, H.; Yamanaka, K. On the Large-signal Modeling of High Power AlGaN/GaN HEMTs. In Proceedings of the 2012 IEEE/MTT-S International Microwave Symposium Digest, Montreal, QC, Canada, 17–22 June 2012.

18. Radhakrishna, U.; Choi, P.; Grajal, J.; Peh, L.; Palacios, T.; Antoniadis, D. Study of RF-circuit Linearity Performance of GAN HEMT Technology Using the MVSG Compact Device Model. In Proceedings of the 2016 IEEE International Electron Devices Meeting, San Francisco, CA, USA, 3–7 December 2016.

19. Ahsan, S.A.; Ghosh, S.; Khandelwal, S.; Chauhan, Y.S. Physics-based multi-bias RF large-signal GaN HEMT modeling and parameter extraction flow. *IEEE J. Electron Devices Soc.* **2017**, *5*, 310–319. [CrossRef]

20. Mao, S.; Xu, Y.; Chen, Y.; Fu, W.; Zhao, X.; Xu, R. High frequency noise model of ALGaN/GaN HEMTs. *ECS J. Solid State Sci. Technol.* **2017**, *6*, S3072–S3077. [CrossRef]

21. Kuang, J.B.; Tasker, P.J.; Wang, G.W.; Chen, Y.K.; Eastman, L.F.; Aina, O.A.; Hier, H.; Fathimulla, A. Kink effect in submicrometer-gate MBE-grown InAlAs/InGaAs/InAlAs heterojunction MESFETs. *IEEE Electron Device Lett.* **1988**, *9*, 630–632. [CrossRef]

22. Meneghesso, G.; Zanon, F.; Uren, M.J.; Zanoni, E. Anomalous kink effect in GaN high electron mobility transistors. *IEEE Electron Device Lett.* **2009**, *30*, 100–102. [CrossRef]

23. Wang, M.; Chen, K.J. Kink effect in AlGaN/GaN HEMTs induced by drain and gate pumping. *IEEE Electron Device Lett.* **2011**, *32*, 482–484. [CrossRef]

24. Crupi, G.; Raffo, A.; Marinković, Z.; Avolio, G.; Caddemi, A.; Marković, V.; Vannini, G.; Schreurs, D.M.M. An extensive experimental analysis of the kink effects in S_{22} and h_{21} for a GaN HEMT. *IEEE Trans. Microw. Theory Tech.* **2014**, *62*, 513–520. [CrossRef]

25. Birafane, A.; Aflaki, P.; Kouki, A.B.; Ghannouchi, F.M. Enhanced DC model for GaN HEMT transistors with built-in thermal and trapping effects. *Solid State Electron.* **2012**, *76*, 77–83. [CrossRef]

26. Cuerdo, R.; Pei, Y.; Chen, Z.; Keller, S.; DenBaars, S.P.; Calle, F.; Mishra, U.K. The kink effect at cryogenic temperatures in deep submicron AlGaN/GaN HEMTs. *IEEE Electron Device Lett.* **2009**, *30*, 209–212. [CrossRef]

27. Fu, L.; Lu, H.; Chen, D.; Zhang, R.; Zheng, Y.; Chen, T.; Wei, K.; Liu, X. Field-dependent carrier trapping induced kink effect in AlGaN/GaN high electron mobility transistors. *Appl. Phys. Lett.* **2011**, *98*, 586. [CrossRef]

28. Brunel, L.; Malbert, N.; Curutchet, A.; Labat, N.; Lambert, B. Kink effect characterization in AlGaN/GaN HEMTs by DC and drain current transient measurements. In Proceedings of the 2012 European Solid-State Device Research Conference (ESSDERC), Bordeaux, France, 17–21 September 2012; pp. 270–273. [CrossRef]

29. Xu, Y.; Wang, C.; Sun, H.; Wen, Z.; Wu, Y.; Xu, R.; Yu, X.; Ren, C.; Wang, Z.; Zhang, B.; Chen, T.; Gao, T. A scalable large-signal multiharmonic model of AlGaN/GaN HEMTs and its application in C-band high power amplifier MMIC. *IEEE Trans. Microw. Theory Tech.* **2017**, *65*, 2836–2846. [CrossRef]

30. Wang, C.; Xu, Y.; Yu, X.; Ren, C.; Wang, Z.; Lu, H.; Chen, T.; Zhang, B.; Xu, R. An electrothermal model for empirical large- signal modeling of AlGaN/GaN HEMTs including self-heating and ambient temperature effects. *IEEE Trans. Microw. Theory Tech.* **2014**, *62*, 2878–2887. [CrossRef]

micromachines

MDPI

Article

An Improved UU-MESFET with High Power Added Efficiency

Hujun Jia *, Mei Hu and Shunwei Zhu

Key Laboratory of the Ministry of Education for Wide Band-Gap Semiconductor Materials and Devices,
School of Microelectronics, Xidian University, Xi'an 710071, China; humei@stu.xidian.edu.cn (M.H.);
swzhu@stu.xidian.edu.cn (S.Z.)
* Correspondence: hjjia@mail.xidian.edu.cn; Tel.: +86-137-7212-6387

Received: 25 September 2018; Accepted: 30 October 2018; Published: 5 November 2018

check for
updates

Abstract: An improved ultrahigh upper gate 4H-SiC metal semiconductor field effect transistor (IUU-MESFET) is proposed in this paper. The structure is obtained by modifying the ultrahigh upper gate height h of the ultrahigh upper gate 4H-SiC metal semiconductor field effect transistor (UU-MESFET) structure, and the h is 0.1 μm and 0.2 μm for the IUU-MESFET and UU-MESFET, respectively. Compared with the UU-MESFET, the IUU-MESFET structure has a greater threshold voltage and trans-conductance, and smaller breakdown voltage and saturation drain current, and when the ultrahigh upper gate height h is 0.1 μm, the relationship between these parameters is balanced, so as to solve the contradictory relationship that these parameters cannot be improved simultaneously. Therefore, the power added efficiency (PAE) of the IUU-MESFET structure is increased from 60.16% to 70.99% compared with the UU-MESFET, and advanced by 18%.

Keywords: 4H-SiC; MESFET; ultrahigh upper gate height; power added efficiency

---

1. Introduction

As a representative of the third generation semiconductor power radio frequency (RF) device, 4H-SiC metal semiconductor field effect transistors (MESFETs) have excellent DC and RF characteristics, such as a high output power density, large saturation current, high breakdown voltage, and large trans-conductance [1–6]. Therefore, 4H-SiC MESFETs have great potential in the fields of radars, electronic countermeasures, and other electronic systems. In recent years, many scholars have devoted themselves to studying the direct-current (DC) and RF characteristics of 4H-SiC MESFETs to meet the requirements of the development of electronic science and technology for 4H-SiC MESFETs. However, in response to the national "Energy Conservation and Emission Reduction, the Green Development" call [7,8], to improve the power added efficiency (PAE) of 4H-SiC MESFETs will become a new trend of research and development.

In this paper, an improved ultrahigh upper gate 4H-SiC metal semiconductor field effect transistor (IUU-MESFET) structure with high PAE is proposed based on an ultrahigh upper gate 4H-SiC metal semiconductor field effect transistor (UU-MESFET) [9]. The proposed IUU-MESFET structure achieves high PAE by modifying the ultrahigh upper gate height h of the UU-MESFET. This is because increasing the ultrahigh upper gate height h can reduce the area of the depletion layer under the ultrahigh upper gate, which not only increases the saturated drain current, but also restrains the expansion of the depletion layer to source/drain sides and reduces the gate-source capacitance. Meanwhile, the enhancement of the ultrahigh upper gate height h alleviates the edge effect of the electric field, thereby improving the breakdown voltage. In sum, the ultrahigh upper gate h affects the DC and RF characteristics of the device, affecting the PAE of the structure. The IUU-MESFET structure has a larger threshold voltage and trans-conductance, and smaller breakdown voltage and saturated drain current, compared with the UU-MESFET. Additionally, a small threshold voltage absolute value indicates that the IUU-MESFET device is more easily depleted from the steering inversion; a high trans-conductance illustrates that the decrease of the distance between the ultrahigh upper gate and the bottom of the channel makes the gate voltage more capable of controlling the drain current in the channel. In general, the IUU-MESFET balances the size relationship between the structure parameters, so that the structure has high PAE and better DC and RF characteristics.

2. Device Structure

The schematic cross sections of the UU-MESFET and IUU-MESFET structures are shown in Figure 1a,b, respectively. From the bottom to the top of the two structures are, in order, a semi-insulating substrate, a p type buffer with a doping concentration of 1.4×10^{15} cm^{-3} and a thickness of 0.5 μm, an n type channel with a doping concentration of 3×10^{17} cm^{-3} and a thickness of 0.25 μm, and two highly doped n type cap layers with a doping concentration of 1×10^{20} cm^{-3} and a thickness of 0.2 μm. The same dimensions are as follows: gate-source spacing L_{gs} = 0.5 μm, gate length L_g = 0.7 μm, gate-drain spacing L_{gd} = 1 μm, the low gate length is 0.35 μm, and the channel is etched 0.05 μm to form the low gate. However, the obvious difference between the UU-MESFET and IUU-MESFET is the ultrahigh upper gate height h. The h of the two structures is 0.2 μm and 0.1 μm, respectively.

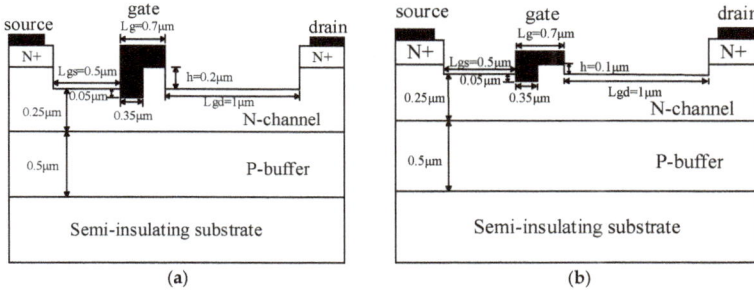

Figure 1. (a) Structural cross sections of ultrahigh upper gate 4H-SiC metal semiconductor field effect transistor (UU-MESFET); (b) Structural cross sections of improved ultrahigh upper gate 4H-SiC metal semiconductor field effect transistor (IUU-MESFET).

The DC and RF characteristics of the two structures are simulated by the two-dimensional simulation software integrated systems engineering technology computer aided design (ISE-TCAD) based on three basic equations of semiconductors (Poisson equation, electron and hole continuity equation, and electron and hole drift and diffusion equation). In the process of advanced design system (ADS) simulation [10], the eesof scalable nonliear GaAsFet model (EE_FET3) is used because the model satisfies the characteristics of 4H-SiC MESFETs. The modified EE_FET3 model is put into the "Load Pull-PAE, Output Power Contours" of ADS for simulation. Additionally, the working conditions are set as follows: V_{gs} is 3.2 V, V_{ds} is 28 V, RF_Req is 850 MHz, Pavs_dBm is 24 dBm, and characteristic impedance Z_0 is 50 Ω. The influence of parameters on PAE is obtained by changing the structural parameters of the device and maintaining the working conditions.

3. Results and Discussion

3.1. The Influence of Structural Parameters on Power Added Efficiency (PAE)

Figures 2 and 3 show the changes in PAE with the trans-conductance (g_m), the saturation drain current (I_d), the breakdown voltage (V_b), and the threshold voltage (V_t) for the UU-MESFET structure. It is found that improving the trans-conductance, the saturation drain current, the breakdown voltage, and the forward conduction threshold voltage is beneficial to increasing the PAE of the UU-MESFET structure. By comparing the influence degree of structural parameters on PAE, it can be seen that the threshold voltage has the greatest impact on PAE, followed by trans-conductance and the breakdown voltage, and the saturation drain current is the least affected. However, there is a contradiction between the structure parameters of the IUU-MESFET device, and the structure parameters cannot be added at the same time, so a suitable ultrahigh upper gate height *h* is needed to balance these parameters so as to obtain a higher PAE.

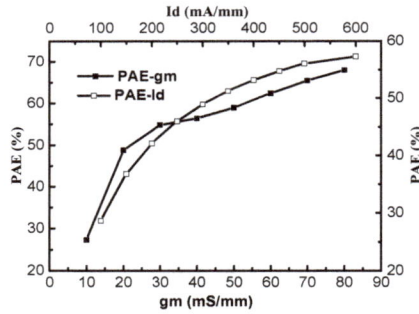

Figure 2. The effect of structural parameters g_m and I_d on power added efficiency (PAE).

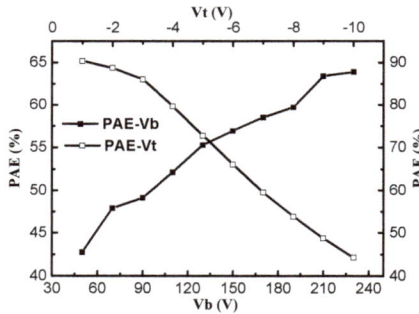

Figure 3. The effect of structural parameters V_b and V_t on PAE.

PAE represents the power amplification capability of the device [10]. Its mathematical expression is shown as (1). Furthermore, the maximum output power density for a Class A amplifier is given by expression (2).

$$\text{PAE} = \frac{P_{out} - P_{in}}{P_{dc}} \tag{1}$$

$$P_{max} = \frac{I_d(V_b - V_{knee})}{8} \tag{2}$$

where P_{out} is the output power of the device, P_{in} is the input power, P_{dc} is the DC dissipative power, and V_{knee} is the knee voltage. The combination of expressions (1) and (2) can prove that increasing the saturation drain current and the breakdown voltage of the device can enhance the PAE. However, the mechanism of increasing PAE by raising the forward conduction threshold voltage and trans-conductance remains to be explored.

3.2. Optimization and Analysis of the Device Structure

Figure 4a,b show the optimization results obtained by changing the ultrahigh upper gate height *h*. It can be seen that the device has the highest PAE when the ultrahigh upper gate height *h* is 0.1 μm, and with the increase of the ultrahigh upper gate height *h*, the saturation drain current and breakdown voltage first increase, and finally tend to saturate, whereas the trans-conductance and threshold voltage decrease by degrees. Therefore, it is impossible to improve PAE by increasing the breakdown voltage, threshold voltage, saturation drain current, and trans-conductance at the same time.

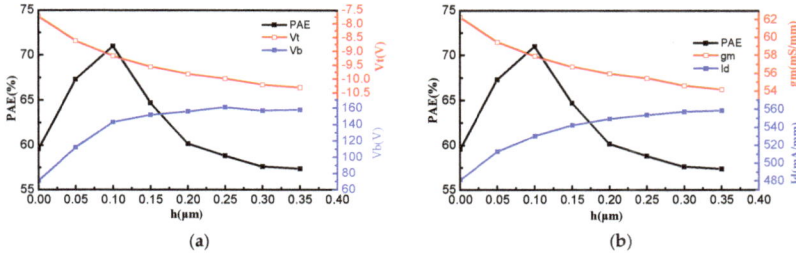

Figure 4. (**a**) The effect of the ultrahigh upper gate height h on PAE, V_t and V_b; (**b**) The effect of the ultrahigh upper gate height h on PAE, V_t and V_b.

In order to solve the problem that the structural parameters cannot be increased simultaneously, it is necessary to find a suitable ultrahigh upper gate height h to balance the relationship between these structure parameters, so as to obtain a larger PAE. As can be seen from Figure 4, the suitable ultrahigh upper gate height h is 0.1 μm. When the ultrahigh upper gate height h is 0.1 μm, the device structure has a larger threshold voltage and trans-conductance, and smaller breakdown voltage and saturation drain current, compared with the device structure with an ultrahigh upper gate height of 0.2 μm. Hence, in order to improve the PAE of the UU-MESFET, we must balance the relationship between structural parameters while pursuing the increase of breakdown voltage, threshold voltage, saturation drain current, and trans-conductance.

3.3. Discussion of the Device Structure

It can be seen from Figure 4 that the optimal structure known as the IUU-MESFET device is obtained when the ultrahigh upper gate height h = 0.1 μm. Table 1 shows the simulation results by ISE TCAD and ADS for the UU-MESFET and IUU-MESFET. According to the Table 1, the PAE values of the two structures are 60.16% and 70.99%, respectively. It can be calculated that the IUU-MESFET structure has an approximately 18% larger PAE value than that of the UU-MESFET structure. Therefore, the IUU-MESFET structure obtains a significant improvement in PAE.

Table 1. Comparison of structural parameters for the two structures.

Parameter	UU-MESFET	IUU-MESFET
V_t (V)	−9.82	−9.17
g_m (mS/mm)	55.93	57.84
V_b (V)	156	143
I_d (mA/mm)	549.28	530.20
PAE (%)	60.16	70.99

It can also be found from Table 1 that the UU-MESFET structure has a larger saturation drain current and breakdown voltage, and the IUU-MESFET structure has a greater threshold voltage and trans-conductance. Therefore, from the DC and RF characteristics of the MESFETs, the UU-MESFET structure is a good choice. However, in terms of efficiency, the UU-MESFET structure is the best. This shows that PAE is not always the best when the MESFETs exhibit a good performance.

4. Conclusions

In this paper, how to improve the PAE of the UU-MESFET structure has been studied, and the IUU-MESFET structure with high PAE is obtained when the ultrahigh upper gate height h is 0.1 μm. The simulation results indicate that improving the threshold voltage, trans-conductance, breakdown voltage, and saturation drain current can increase the PAE of the UU-MESFET. It is found that enhancing the ultrahigh upper gate height h can increase the breakdown voltage and saturation drain current, and reduce the trans-conductance and threshold voltage. When h is 0.1 μm, the IUU-MESFET structure has a high PAE value and better DC and RF characteristics. The simulation results of ISE TCAD and ADS show that in order to improve the PAE of the UU-MESFET device, the breakdown voltage, the threshold voltage, the trans-conductance, and the saturation drain current should be increased, and the size relationship between them should be balanced. This may serve as a general design direction for improving the PAE of 4H-SiC MESFETs.

Author Contributions: Project Administration, H.J.; Writing—Original Draft Preparation, M.H.; Writing—Review and Editing, S.Z.

Funding: This work was supported in part by the National Natural Science Foundation of China (NSFC) under Grant No. 61671343, and in part by the National Key Basic Research Program of China (973 Program) under grant No. 2014CB339900.

Conflicts of Interest: The authors declare no conflict of interest.

References

1. Cha, H.Y.; Thomas, C.I.; Koley, G.; Eastman, L.F.; Spencer, M.G. Reduced trapping effects and improved electrical performance in buried-gate 4H-SiC MESFETs. *IEEE Trans. Electron Devices* **2003**, *50*, 1569–1574.
2. Zhu, C.L.; Rusli; Tin, C.C.; Zhang, G.H.; Yoon, S.F.; Ahn, J. Improved performance of SiC MESFETs using double-recessed structure. *Microelectron. Eng.* **2006**, *83*, 92–95. [CrossRef]
3. Andersson, K.; Sudow, M.; Nilsson, P.A.; Sveinbjornsson, E.; Hjelmgren, H.; Nilsson, J.; Stahl, J.; Zirath, H.; Rorsman, N. Fabrication and characterization of field-plated buried-gate SiC MESFETs. *IEEE Electron Device Lett.* **2006**, *27*, 573–575. [CrossRef]
4. Jia, H.J.; Yang, Z.H.; Wu, Q.Y.; Ma, P.M. Improved double-recessed P-buffer 4H-SiC MESFETs with partial heavy doped channel. *Mater. Sci. Semicond. Process.* **2016**, *56*, 213–216. [CrossRef]
5. Zhang, J.P.; Luo, X.R.; Li, Z.J.; Zhang, B. Improved double-recessed 4H-SiC MESFETs structure with recessed source/drain drift region. *Microelectron. Eng.* **2007**, *84*, 2888–2891. [CrossRef]
6. Jia, H.J.; Zhang, H.; Yang, Y.T. A novel 4H-SiC MESFET with an L-gate and a partial p-type spacer. *Mater. Sci. Semicond. Process.* **2012**, *15*, 2–5. [CrossRef]
7. Available online: http://www.xinhuanet.com/politics/2016lh/2016-03/17/c_1118366322.htm (accessed on 5 November 2018).
8. Available online: http://www.shcpo.com.cn/index.php/law/gjbwwj/245-2016-2020 (accessed on 5 November 2018).
9. Jia, H.J.; Zhang, H.; Xing, D.; Luo, Y.H.; Duan, B.X. A novel 4H-SiC MESFET with ultrahigh upper gate. *Superlattices Microstruct.* **2015**, *86*, 372–378. [CrossRef]
10. Cho, K.J.; Kim, J.H.; Stapleton, S.P. RF high power Doherty amplifier for improving the efficiency of a Feedforward linear amplifier. In Proceedings of the 2004 IEEE MTT-S International Microwave Symposium Digest, Fort Worth, TX, USA, 6–11 June 2004.

micromachines

MDPI

Article

A Novel One-Transistor Dynamic Random-Access Memory (1T DRAM) Featuring Partially Inserted Wide-Bandgap Double Barriers for High-Temperature Applications

Myeongsun Kim [1], Jongmin Ha [1], Ikhyeon Kwon [2], Jae-Hee Han [3], Seongjae Cho [2,*] and Il Hwan Cho [1,*]

[1]   Department of Electronic Engineering, Myongji University, Yongin-si, Gyeonggi-do 17058, Korea;
      kimms0700@gmail.com (M.K.); hjm2703@naver.com (J.H.)
[2]   Department of IT Convergence Engineering, Gachon University, Seongnam-si, Gyeonggi-do 13120, Korea;
      kih3596@gmail.com
[3]   Department of Energy IT, Gachon University, Seongnam-si, Gyeonggi-do 13120, Korea;
      jhhan388@gachon.ac.kr
*   Correspondence: felixcho@gachon.ac.kr (S.C.); ihcho77@mju.ac.kr (I.H.C.)

Received: 1 October 2018; Accepted: 5 November 2018; Published: 7 November 2018

Abstract: These days, the demand on electronic systems operating at high temperature is increasing owing to bursting interest in applications adaptable to harsh environments on earth, as well as in the unpaved spaces in the universe. However, research on memory technologies suitable to high-temperature conditions have been seldom reported yet. In this work, a novel one-transistor dynamic random-access memory (1T DRAM) featuring the device channel with partially inserted wide-bandgap semiconductor material toward the high-temperature application is proposed and designed, and its device performances are investigated with an emphasis at 500 K. The possibilities of the program operation by impact ionization and the erase operation via drift conduction by a properly high drain voltage have been verified through a series of technology computer-aided design (TCAD) device simulations at 500 K. Analyses of the energy-band structures in the hold state reveals that the electrons stored in the channel can be effectively confined and retained by the surrounding thin wide-bandgap semiconductor barriers. Additionally, for more realistic and practical claims, transient characteristics of the proposed volatile memory device have been closely investigated quantifying the programming window and retention time. Although there is an inevitable degradation in state-1/state-0 current ratio compared with the case of room-temperature operation, the high-temperature operation capabilities of the proposed memory device at 500 K have been confirmed to fall into the regime permissible for practical use.

Keywords: harsh environment; space application; 1T DRAM; wide-bandgap semiconductor; high-temperature operation; TCAD

1. Introduction

An integrated electronic system capable of operating at high temperature would be beneficial to various industrial applications, harsh environment systems, and core functional components for the aerospace systems [1,2]. When a semiconductor device is operated in a high-temperature environment, a number of problems mainly caused by the leakage currents due to greatly increased generation rate of electron-hole pairs (EHPs) are more likely to take place [3]. In order to resolve the issues, wide-bandgap materials such as GaN have been usually employed as the platform for the applications, instead of Si [4,5], by which the number of thermally generated carriers threatening the ideal electronic device performances can be reduced. Although there have been studies on high-temperature-operation transistors based on wide-bandgap materials towards the purpose, relatively less interest has been devoted to high-temperature memory technology. Additionally, although there is some research on nonvolatile memories [6–8], high-speed volatile memories coping with the processing unit in the specifically designed system have great deal of room to delve into.

In this work, we develop a novel volatile memory having high-temperature operation capabilities. In most of the conventional one-transistor dynamic random-access memory (1T DRAM), the holes stored in the channel region modulate the threshold voltage and the drain current level in performing the read operation [9]. Additionally, in the conventional 1T DRAM device, the hole storage is provided by the energy barriers constructed by PN junctions at both ends of the channel and by the buried oxide (BOX). The stored holes can have several leakage paths, such as recombination, drift, diffusion, and inter-band tunneling [3]. 1T DRAMs in various novel structures have been proposed to suppress the data leakage and increase the retention time [10–12]. However, the previous studies are limited to room-temperature operation in most cases. An existing study introduces a 1T DRAM operating at high temperature, but the confirmed upper limit is 370 K and the subsequent studies are not active yet [13]. In this work, we propose, design, and characterize a novel 1T DRAM featuring a physical barrier made of wide-bandgap semiconductor material which confines the stored carriers highly effectively, verifying the memory operations through series of technology computer-aided design (TCAD) simulation works.

2. Device Structure and Simulation Strategy

The proposed 1T DRAM device with a pair of partially introduced vertical thin wide-bandgap barriers is illustrated in Figure 1a. As mentioned briefly, as temperature increases, junction leakage increases, and the stored holes smear out of the storage in most 1T DRAM devices [14]. Although the energy barrier formed by the gate oxide is as high as 3 eV, that introduced by a Si PN junction is at most the bandgap energy of Si, approximately 1.12 eV. Therefore, the leakage through source/drain junction becomes more prominent as temperature gets higher, owing to the carriers occupying the tail states of the Fermi-Dirac distribution with higher flatness about the Fermi level due to the temperature effect. In order to reduce the leakage currents stemming from the carriers coming over the energy barrier by PN junction, the 1T DRAM proposed in this work employs wide-bandgap material on both borders between channel and source/drain junctions. GaP has the least lattice mismatching among the single-species and compound semiconductors that can be used for device fabrication. We have sought the wide-bandgap materials with the highest degree of lattice matching for this application [15]. Unfortunately, GaP processing is not allowed in most of the Si CMOS clean rooms yet but molecular beam epitaxy (MBE) or metal-organic chemical vapor deposition (MOCVD) needs to be schemed for the epitaxial growth of GaP on Si. Si and GaP have similar thermal expansion coefficients of 2.6×10^{-6} K^{-1} and 4.65×10^{-6} K^{-1}, respectively, at 300 K. Moreover, the thermal expansion coefficient of Si monotonically increases with temperature and that of GaP increases but shows a relatively slow slope, which makes both of them practically the same at the processing temperature above 800 K. Thus, the thermal expansion coefficient matching acts as another merit in forming the Si/GaP heterostructure. A more tangible effect of lattice matching and thermal expansion coefficient matching can be quantified as the interface trap density eventually. The interface trap densities between Si and GaP and between GaP and SiO$_2$ were reported to be 1×10^{13} cm^{-2} and 7×10^{12} cm^{-2}, respectively [16,17], which are comparably low as the trap density between Si and SiGe under a well-controlled epitaxial growth. This favorable interface status between Si and GaP results in the permissibly low off-state current in the metal-oxide-semiconductor field-effect transistor (MOSFET) operation. The wide-bandgap of GaP provides the energy barriers at both ends of the Si channel, which more effectively confines the carriers at high temperature compared with the energy barriers electrically formed by counter-doped Si regions [18]. SiC can be also adopted for our application in the sense that SiC has a stronger Si processing compatibility and wider energy bandgap than GaP. There are several different phases of SiC, but all the bandgap energies are larger than that of GaP, 2.26 eV. Thus, more effective carrier confinements become presumable with SiC for its application to high-temperature 1T DRAM technology. On the other hand, the lattice mismatch between Si and SiC is larger than that between Si and GaP regardless of the phases of SiC. Accurate control of the SiC barrier thickness should be performed in consideration of its epitaxial critical thickness on Si for being more affirmative with its application.

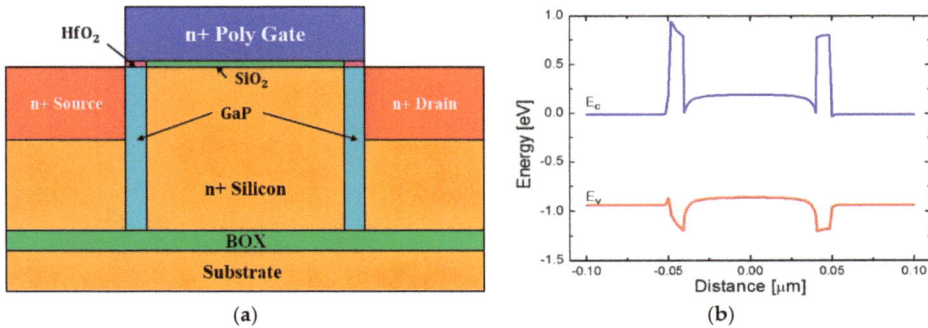

Figure 1. Device structure. (**a**) Schematic of the one-transistor dynamic random-access memory (1T DRAM) with partially introduced wide-bandgap barriers; (**b**) energy-band diagram along the channel direction in the proposed device.

Figure 1b shows the energy-band diagram along the channel direction beneath the gate oxide. As can be confirmed by Figure 1b, since the large difference between bandgap energies of Si and GaP is mostly projected to the conduction band offset (CBO), the energy barrier seen by the conduction electrons is considerably higher than the barrier in the valence band, valence band offset (VBO), seen by the conduction holes. This high energy barrier in the conduction band effectively prevents the stored electrons from escaping to either source or drain junction, even at an elevated temperature. In order to make full use of the beneficial energy-band structure, the proposed 1T DRAM device stores electrons in a way different from that employed by most of the previously reported 1T DRAM, having a n^+ channel to minimize unwanted loss of stored electrons by recombination in preserving the stored data. The electrical characteristics and the memory operations of the proposed 1T DRAM cell have been investigated by a commercial TCAD package, Sentaurus by Synopsys (Mountain View, CA, USA). Gate length is 100 nm, thicknesses of gate oxide and HfO_2 are both 3 nm, thickness of BOX is 10 nm, and barrier width and depth are designed to be 10 nm and 75 nm, respectively. The doping concentrations of source/drain junctions and substrate are n-type 1×10^{20} cm^{-3} and n-type 1×10^{16} cm^{-3}, respectively. A number of physical models including the Shockley-Read-Hall recombination model, Fermi statistics model, band-to-band tunneling model, and doping- and electric field-dependent mobility models are activated simultaneously in cooperation for higher accuracy and reliability of the simulation results. In particular, temperature models have been employed for reflecting the temperature effects. The highest operating temperature of a commercial memory device is known to be about 400 K. Considering the temperature robustness of metals used in the back-end-of-the-line (BEOL), the characterization and evaluation have been carried out at 500 K, which far extends the known upper limit of temperature warranting the permissible memory operations.

3. Results

In typical MOSFET transistors and barrier-assisted 1T DRAMs, the energy barriers formed between the channel and the source/drain junctions control the off-state current. The role is taken over by the Si/GaP heterojunction in the proposed 1T DRAM device. Although the barrier introduced by the GaP heterojunction is even larger than that by the Si anisotype homojunction, the energy barrier can be effectively lowered by a high enough gate voltage. The tiny segments of HfO_2 at both ends of the gate oxide are positioned to enhance the gate controllability over the barrier height of GaP, which is required for program operation which draws electrons from outside into the channel.

Figure 2a shows the I_D-V_{GS} characteristic curves of the proposed 1T DRAM device at different temperatures of 300 K, 400 K, and 500 K. The current increases in both low and high V_{GS} regimes, which is owing to the increase of thermally generated carriers. Figure 2b demonstrates the output characteristic curves at different V_{GS} values. Operation voltage is higher than that of Si MOSFET with a comparable channel length since V_{GS} needs to be high enough to lower the high energy barrier brought by GaP for electron conduction. Focus in this work is made on the high-temperature operation capabilities of the proposed device eventually aiming the applications in the extremely harsh environment, and the dimension and drive voltage scaling criteria have not been taken forward. As can be confirmed by comparing Figures 1b and 2c, applying a high enough V_{GS} lowers the energy barrier between the channel and the source/drain junctions. Additionally, since the electrons see an increase number of allowed energy states by the increased occupation probability determined by the Fermi-Dirac distribution at 500 K, a significant number of carriers overcome the energy barrier even though the higher energy barrier is constructed compared with the case of previously reported 1T DRAMs [3].

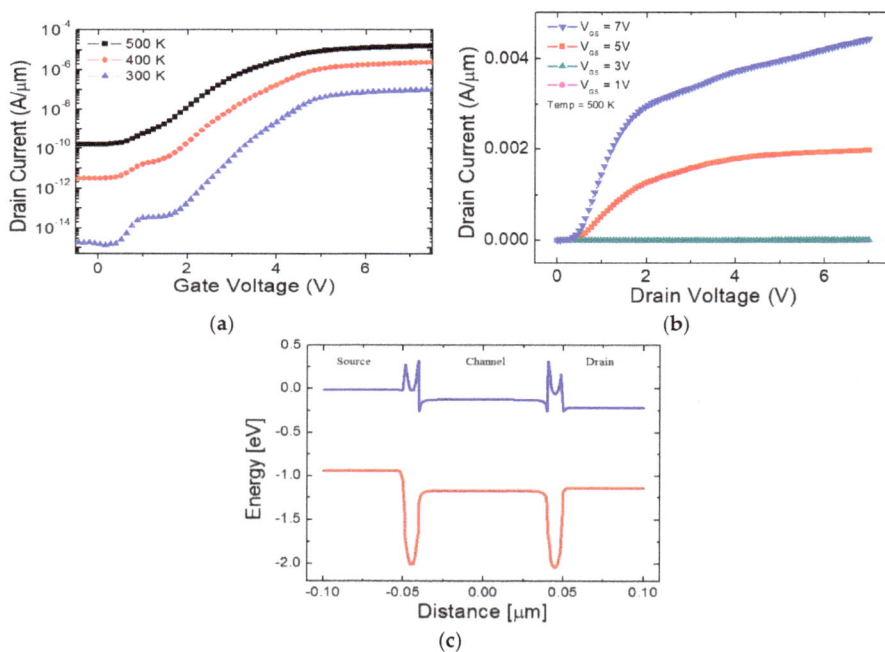

Figure 2. Operation characteristics. (**a**) I_D-V_{GS} characteristic curves with V_{DS} = 0.2 V at 300 K, 400 K, and 500 K; (**b**) I_D-V_{DS} characteristic curves at different V_{GS} values at 500 K; (**c**) energy-band diagram along the channel beneath the gate oxide at V_{GS} = 8 V and V_{DS} = 0.2 V at 500 K.

3.1. Program and Erase Operation Schemes

In order for the proposed device to operate as a 1T DRAM, it is necessary to create and store carriers to change the threshold voltage of the device. Electrons are created by impact ionization, which is one of the conventional carrier generation methods in 1T DRAM devices [19]. The proposed device differs from the existing ones in the sense that electrons are used instead of holes between carriers generated by impact ionization. As mentioned above, the energy barrier for the electrons constructed by Si/GaP heterostructure is larger than that for holes. In the erase operation, a large positive voltage is applied to the drain and a small negative voltage is applied to the gate. The detailed bias conditions for all the memory operations are summarized in Table 1.

Table 1. Bias conditions for program, erase, and hold operations.

Program		Erase		Read		Hold	
V_{GS}	V_{DS}	V_{GS}	V_{DS}	V_{GS}	V_{DS}	V_{GS}	V_{DS}
−0.5 V	−5.0 V	−0.5 V	5.0 V	0.5 V	0.2 V	0 V	0 V

In order to give rise to impact ionization in the program operation, a large electric field should be applied in the channel direction. Figure 3a shows the simulated energy-band diagram under the high lateral electric field. The influences of the large-magnitude negative drain voltage and the small negative gate voltage are combined so that the energy barrier due to GaP substantially disappears. As the energy barrier introduced by GaP is lowered, it becomes possible to inject a large number of carriers into the channel while performing the program operation.

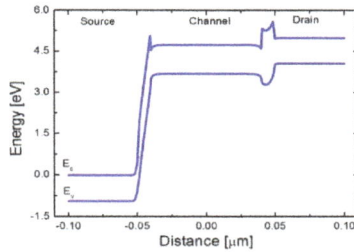

(a)

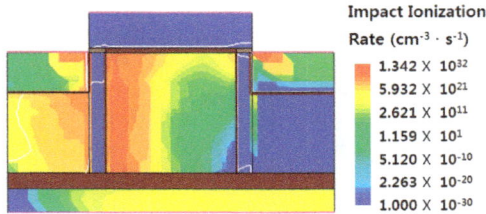

(b)

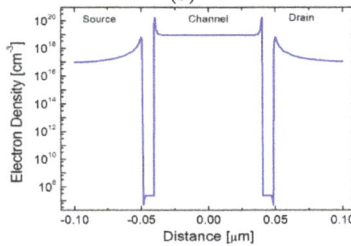

(c)

Figure 3. Verification of program operation. (**a**) Energy-band diagram along the channel direction under program bias condition at 500 K; (**b**) contour of impact ionization rate in the program operation at 500 K; (**c**) electron concentration after program operation at 500 K.

The program operation can be confirmed by investigating the contour of the impact ionization rate over the device as demonstrated in Figure 3b. The high impact ionization rate is observed near the Si/GaP heterostructure near the source junction. The change in concentration of electrons stored in the channel after a program operation is depicted in Figure 3c. Electron concentration as high as 1×10^{19} cm^{-3} in the channel storage is assured even in the high-temperature environment at 500 K.

For removing the stored electrons, a large positive voltage is applied to the drain junction. As shown in Figure 4a, since the energy barrier between the channel and the drain gets lower, the electrons in the channel are repelled to the drain junction. Additionally, a small negative gate voltage assists the repelling force on the electrons stored in the channel. As a result, the electron concentration in the channel is reduced as shown in Figure 4b, which can be clearly confirmed by comparison with Figure 3c. Since a massive amount of EHPs are generated at 500 K, even the lowest concentration of electrons in the channel region is higher than the channel doping concentration. The electron concentration increase by the EHPs degrades the read current ratio at the elevated temperature.

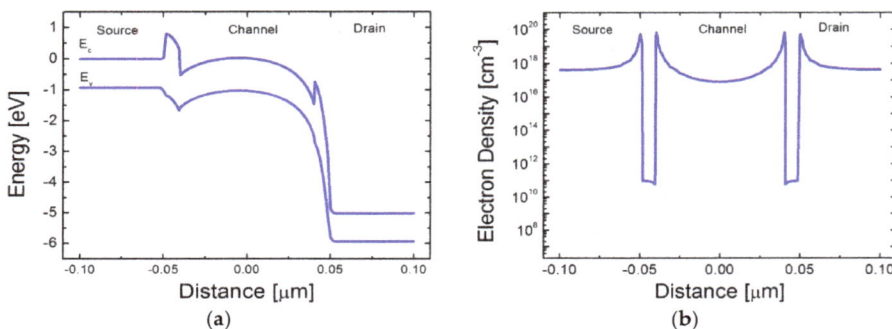

Figure 4. Verification of erase operation. (**a**) Energy-band diagram along the channel direction under the erase bias condition at 500 K; (**b**) Electron concentration after erase operation at 500 K.

3.2. Hold Operation and Retention Characteristics

In the 1T DRAM technology, hold operation can be optionally prepared to maintain the number of stored carriers, after program and erase operations before the read operation. For reducing the total power consumption over a period of memory operations, zero voltages are applied to both gate and drain terminals for holding the carriers stored in the channel.

Figure 5a,b shows the potential distribution and the energy-band diagram under the hold-1 and hold-0 conditions, respectively. In state 1, electrons are generated and the potential in the channel region is lowered. On the other hand, in state 0, the electrons are removed, and the potential is recovered to a high value leveling with the potentials in the source and drain junctions. Figure 5c depicts the energy-band diagrams in the channel direction in state 1 and 0 for the explicit comparison. The most important feature in the high-temperature operations is determined by whether a sufficiently high energy barrier can be steered or not so that the stored electrons are not allowed to escape to either source or drain junctions in state 1 at a high temperature. The original high electron energy in Figure 1b gets lower as electrons are accumulated in the channel as the program operation is progressed. However, as can be seen in Figure 5c, a high electron potential energy barrier of 0.5 eV still exists between the channel and the source/drain junctions in state 1. The energy barrier can be changed by introducing different wide-bandgap semiconductor material in the physical barrier with different energy bandgap and electron affinity.

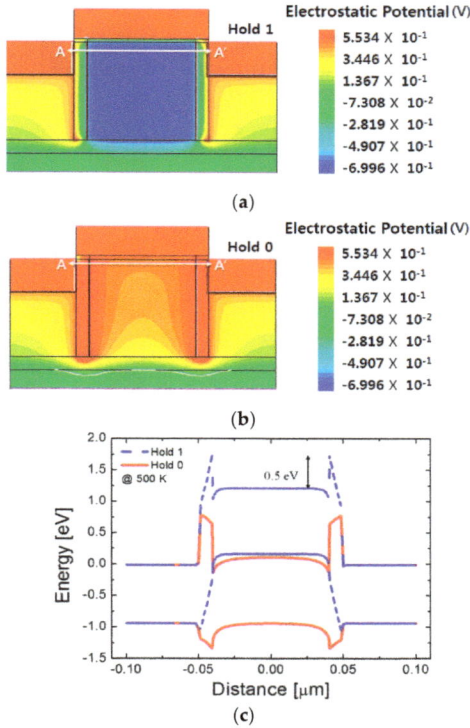

Hold 1

Electrostatic Potential (V)

5.534 X 10⁻¹
3.446 X 10⁻¹
1.367 X 10⁻¹
-7.308 X 10⁻²
-2.819 X 10⁻¹
-4.907 X 10⁻¹
-6.996 X 10⁻¹

(a)

Hold 0

Electrostatic Potential (V)

5.534 X 10⁻¹
3.446 X 10⁻¹
1.367 X 10⁻¹
-7.308 X 10⁻²
-2.819 X 10⁻¹
-4.907 X 10⁻¹
-6.996 X 10⁻¹

(b)

(c)

Figure 5. Verification of hold operation. (**a**) Potential distribution under the hold-1 condition; (**b**) potential distribution under the hold-0 condition; (**c**) energy-band diagrams under the hold-1 and the hold-0 conditions along the A-A' cutline in (**a**).

3.3. Transient Simulation Results for the Cyclic 1T DRAM Operations

Figure 6 demonstrates the transient simulation results for 1-cycle memory operations of the proposed 1T DRAM. The cyclic operation consists of program/hold 1/read 1/hold 1/erase/hold 0/read 0/(hold 0) and the drain currents in each operation have been extracted at 300 K and 500 K. Read-1/read-0 current ratios are 1000 and 100 at 300 K and 500 K, respectively. Although the current ratio is reduced at 500 K compared with that at room temperature, not a small ratio is still preserved, and the ratio can be practically amplified and sensed by the supplemental functions of peripheral circuits.

The reasons that the current ratio decreases at 500 K can be considered to be the following: (1) Carrier recombination rate increases as the operating temperature increases as shown in Figure 7. For maintaining the steady-state carrier concentration, the recombination rate should be equal to the generation rate and the latter is a strong function of temperature. This also leads to reduction of carrier storage time in the end; (2) high temperature increases EHP generation, which results in higher electron concentration. Thus, the initial current level at 500 K should be significantly higher than that at 300 K as can be confirmed by Figure 6 along with the previous work [15]. Contemplating the quantitative analysis in Figure 6, the reason (2) is considered to be dominant. The carrier storage time, i.e., retention time is defined as the time moment when the read-1 and read-0 current ratio reduces down to 10%. Unlike the conventional DRAM where the retention is evaluated by the bitline voltage drop with time, state current reading should be performed for 1T DRAM where the bitline precharge scheme for read operation is not employed. By this method, the storage time is extracted to be 1 μs. At this moment, we have no idea how long or short this retention might be, since there is no concrete standard and requirement for 500 K operation at all yet. However, it is sure that the obtained retention time is much shorter than that of conventional DRAM in the 1-transistor 1-capacitor (1T1C) configuration.

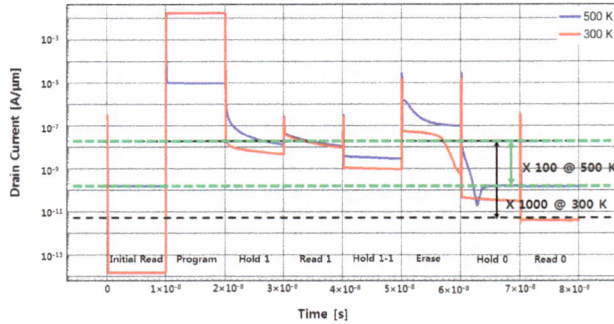

Figure 6. Transient simulation results for memory operations of the proposed 1T DRAM at 300 K and 500 K.

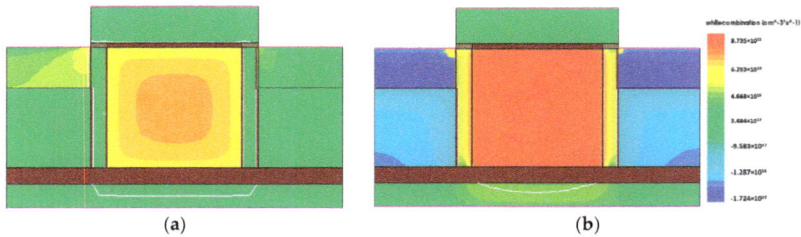

Figure 7. Distributions of the Shockley-Read-Hall (SHR) recombination rates under hold-1 condition at (**a**) 300 K and (**b**) 500 K.

4. Conclusions

We have proposed, designed, and characterized a novel 1T DRAM operational at 500 K featuring double wide-bandgap barriers for elongated data storage capability. The wide-bandgap semiconductor material, GaP, is introduced in the Si platform in order to heighten the energy barrier seen by the electrons stored in the channel storage. The series of simulation results support that the stored electrons in the channel are effectively preserved, even at 500 K. The firstly proposed scheme utilizing Si and lattice-matching wide-bandgap material GaP, and the device structure design, have prepared the strong potential for memory technologies in the high-temperature environment which can be found in the applications for auto-vehicles, industrial turbine systems, and aerospace systems.

Author Contributions: M.K. collaborated on basic ideas and performed the simulations related to device program and erase data. J.H. performed a transient simulation and performed work related to setting the operating voltage. I.K. was responsible for the thermal simulation setup and performed the data sorting and analysis. J.-H.H. gained the previous reports on electron devices for high-temperature applications and thoroughly investigated the possibilities of Si-compatible semiconductor materials in this technology in terms of primary material parameters and process viability. S.C. was responsible for data analysis, physical interpretation of the simulation results, and preparation of evaluation schemes for an unknown 1T DRAM device. I.H.C. jointly proposed the basic idea of this work with S.C., led the research direction, and made guidelines for simulation strategy and data acquisition.

Funding: This work was supported by the Ministry of Trade, Industry and Energy of Korea (MOTIE) and the Korean Semiconductor Research Consortium (Grant No. 10080513) and by the National Research Foundation of Korea (NRF) funded by the Ministry of Science and ICT (MSIT) (Grant No. NRF-2017R1A2B2011570). This work was also supported by Basic Science Research Program through the NRF funded by the Ministry of Education, Science and Technology (Grant No. 2016R1D1A1B03935211) and ICT Program of MSIT/IITP (Grant No. B0101-15-1347).

Micromachines **2018**, *9*, 581

Conflicts of Interest: The authors declare no conflict of interest.

References

1. Elgabra, H.; Siddiqui, A.; Singh, S. Simulation of conventional bipolar logic technologies in 4H-SiC for harsh environment applications. *Jpn. J. Appl. Phys.* **2016**, *55*, 04ER08. [CrossRef]
2. De Jong, P.C.; Meijer, G.C.M.; Van Roermund, A.H.M. A 300 °C dynamic-feedback instrumentation amplifier. *IEEE J. Solid State Circuits* **1998**, *33*, 1999–2009. [CrossRef]
3. Dreike, P.L.; Fleetwood, D.M.; King, D.B.; Sprauer, D.C.; Zipperian, T.E. An overview of high-temperature electronic device technologies and potential applications. *IEEE Trans. Compon. Packag. Manuf. Technol.* **1994**, *17*, 594–609. [CrossRef]
4. Shashikala, B.N.; Nagabhushana, B.S. High temperature performance of wide bandgap semiconductors devices for high power applications. *Int. J. Eng. Sci. Technol.* **2015**, *2*, 7586–7591.
5. Neudeck, P.G.; Okozie, R.S.; Chen, L.Y. High-temperature electronics—A role for wide bandgap semiconductors? *Proc. IEEE* **2002**, *90*, 1065–1076. [CrossRef]
6. Walczyk, C.; Walczyk, D.; Schroeder, T.; Bertaud, T.; Sowinska, M.; Lukosius, M.; Fraschke, M.; Wolansky, D.; Tillack, B.; Miranda, E.; et al. Impact of temperature on the resistive switching behavior of embedded HfO$_2$-based RRAM devices. *IEEE Trans. Electron Devices* **2011**, *58*, 3124–3131. [CrossRef]
7. Cabout, T.; Vianello, E.; Jalaguier, E.; Grampeix, H.; Molas, G.; Blaise, P.; Cueto, O.; Guillermet, M.; Nodin, J.F.; Pemiola, L.; et al. Effect of SET temperature on data retention performances of HfO$_2$-based RRAM cells. In Proceedings of the 2014 IEEE 6th International Memory Workshop (IMW), Taipei, Taiwan, 18–21 May 2014.
8. Kwon, I.; Islam, M.S.; Cho, I.H. Investigation of nonvolatile AlGaN/GaN flash memory for high temperature operation. *J. Semicond. Technol. Sci.* **2018**, *18*, 100–107. [CrossRef]
9. Lin, J.T.; Chang, M. A new 1T DRAM cell with enhanced floating body effect. In Proceedings of the 2006 IEEE International Workshop on Memory Technology, Design, and Testing (MTDT '06), Taipei, Taiwan, 2–4 August 2006.
10. Chen, C.H.; Lin, J.T.; Eng, Y.C.; Lin, P.H.; Chiu, H.N.; Chang, T.F.; Tai, C.H.; Lu, K.Y.; Fan, Y.H.; Chang, Y.C.; et al. A novel vertical MOSFET with bMPI structure for 1T-DRAM application. In Proceedings of the 2010 International Symposium on Next Generation Electronics, Kaohsiung, Taiwan, 18–19 November 2010.
11. Tian, Y.; Huang, R.; Zhang, X.; Wang, Y. A novel nanoscaled device concept: quasi-SOI MOSFET to eliminate the potential weaknesses of UTB SOI MOSFET. *IEEE Trans. Electron Devices* **2005**, *52*, 561–568. [CrossRef]
12. Ertosun, M.G.; Cho, H.; Kapur, P.; Sarswat, K.C. A nanoscale vertical double-gate single-transistor capacitorless DRAM. *IEEE Electron Device Lett.* **2008**, *29*, 615–617. [CrossRef]
13. Lin, J.T.; Lin, P.H. Multifunction behavior of a vertical MOSFET with trench body structure and new erase mechanism for use in 1T-DRAM. *IEEE Trans. Electron Devices* **2014**, *61*, 3172–3178. [CrossRef]
14. Lin, J.T.; Lin, P.H.; Eng, Y.C.; Chen, Y.R. Thermal stability of a vertical SOI-based capacitorless one-transistor DRAM with trench-body structure. *IEEE Trans. Electron Devices* **2013**, *60*, 1872–1877. [CrossRef]
15. Kwon, I.; Kwon, H.I.; Cho, I.H. Development of high temperature operation silicon based MOSFET for harsh environment application. *Results Phys.* **2018**, *11*, 475–481. [CrossRef]
16. Landis, G.A.; Loferski, J.J.; Beaulieu, R.; Sekula-Moise, P.A.; Vernon, S.M.; Spitzer, M.B.; Keavney, C.J. Wide-bandgap epitaxial heterojunction windows for silicon solar cells. *IEEE Trans. Electron Devices* **1990**, *37*, 372–381. [CrossRef]
17. Spitzer, S.M.; Schwartz, B.; Kuhn, M. Electrical properties of a native oxide on gallium phosphide. *J. Electrochem. Soc.* **1973**, *120*, 669–672. [CrossRef]
18. Carnevale, S.D.; Deitz, J.I.; Carlin, J.A.; Picard, Y.N.; McComb, D.W.; De Graef, M.; Ringel, S.A.; Grassman, T.J. Applications of electron channeling contrast imaging for the rapid characterization of extended defects in III–V/Si heterostructures. *IEEE J. Photovolt.* **2015**, *5*, 676–682. [CrossRef]
19. Okhonin, S.; Nagoga, M.; Sallese, J.M.; Fazan, P. A capacitor-less 1T-DRAM cell. *IEEE Electron Device Lett.* **2002**, *23*, 85–87. [CrossRef]

micromachines

MDPI

Article

Influence of Passivation Layers on Positive Gate Bias-Stress Stability of Amorphous InGaZnO Thin-Film Transistors

Yan Zhou and Chengyuan Dong *

Department of Electronic Engineering, Shanghai Jiao Tong University, Shanghai 200240, China;
zhou_yan@sjtu.edu.cn
* Correspondence: cydong@sjtu.edu.cn; Tel.: +86-21-3420-8566

Received: 18 September 2018; Accepted: 14 November 2018; Published: 17 November 2018

check for updates

Abstract: Passivation (PV) layers could effectively improve the positive gate bias-stress (PGBS) stability of amorphous InGaZnO (a-IGZO) thin-film transistors (TFTs), whereas the related physical mechanism remains unclear. In this study, SiO_2 or Al_2O_3 films with different thicknesses were used to passivate the a-IGZO TFTs, making the devices more stable during PGBS tests. With the increase in PV layer thickness, the PGBS stability of a-IGZO TFTs improved due to the stronger barrier effect of the PV layers. When the PV layer thickness was larger than the characteristic length, nearly no threshold voltage shift occurred, indicating that the ambient atmosphere effect rather than the charge trapping dominated the PGBS instability of a-IGZO TFTs in this study. The SiO_2 PV layers showed a better improvement effect than the Al_2O_3 because the former had a smaller characteristic length (~5 nm) than that of the Al_2O_3 PV layers (~10 nm).

Keywords: amorphous InGaZnO (a-IGZO); thin-film transistor (TFT); positive gate bias stress (PGBS); passivation layer; characteristic length

1. Introduction

Amorphous InGaZnO thin-film transistors (a-IGZO TFTs) have considerable potential for applications in next-generation flexible, transparent, and large-size flat panel displays (FPDs) because of their superior electrical characteristics, such as large field-effect mobility (~10 $cm^2/V \cdot s$), low subthreshold swing (~0.2 V/decade), small leakage current (<10^{-13} A), and so on [1,2]. However, the reliability issues, e.g., threshold voltage (V_{th}) shift under positive gate bias stress (PGBS), remain to be solved. Both charge trapping [3–6] and ambient atmosphere effect [7–14] have been reported to be responsible for V_{th} shifts in a-IGZO TFTs during PGBS tests. Meanwhile, some research groups have demonstrated that the bulk trapping effect [15,16] and plasma damage [17,18] could also lead to V_{th} shifts under PGBS. Evidently, this V_{th} instability is not preferred. In fact, PGBS instability may seriously hinder the actual applications of a-IGZO TFTs in FPDs because it may directly impact the brightness uniformity and stability of display panels. Therefore, some measures must be taken to make the devices more stable during PGBS tests. Passivation (PV) layers, such as SiO_2, Si_3N_4, etc., have been reported to exhibit a good resistance to ambient atmosphere, and thus improve the PGBS stability of a-IGZO TFTs [19–24]. However, the exact physical mechanism for how PV layers make devices more stable remains unclear. In this paper, we sputtered SiO_2 (or Al_2O_3) with different thicknesses to passivate a-IGZO TFTs, observing the variation of their PGBS instability. Both SiO_2 and Al_2O_3 were chosen for the PV layers because of their good compatibility in TFT process integration. It was found that the V_{th} shift was reduced by increasing the PV layer thickness and the SiO_2 improved the PGBS stability of a-IGZO TFTs more significantly than the Al_2O_3. The related physical mechanism was classified based on the experimental observations.

2. Materials and Methods

Inverted staggered a-IGZO TFTs were fabricated, the schematic cross-section of which is shown in Figure 1. P-type silicon wafers (gate electrodes) with a 200 nm-thick thermal SiO_2 (gate insulators) were used as substrates. After thorough cleaning, 50 nm-thick a-IGZO films (In:Ga:Zn = 1:1:1 in mol ratio) as the channel layers were prepared on the substrates using radio frequency (RF) magnetron sputtering at room temperature (RT) with a power of 60 W, a pressure of 5 mTorr, and an Ar flow rate of 30 sccm. Then, Indium Tin Oxide (ITO) films with a thickness of 200 nm were deposited as source/drain (S/D) electrodes using direct current (DC) magnetron sputtering at RT, where the power was 100 W, the pressure was 5 mTorr, and the Ar flow rate was 30 sccm. For the passivated devices, SiO_2 (or Al_2O_3) films with different thicknesses were deposited using RF sputtering at RT with ta power of 50 W, a pressure of 5 mTorr, and an Ar flow rate of 30 sccm. The channel layers, S/D electrodes, and PV layers were patterned using shadow masks during their depositions, leading to a channel width/length (W/L) of 1000/275 μm. Finally, the devices were annealed at 400 °C for 1 h.

The electrical characteristics of the TFTs were measured using a 2636 A parameter analyzer (Keithley Instruments, Inc., Beaverton, OR, USA) in an unsealed chamber, which maintained the atmospheric pressure and little gas circulation. The moisture content in the chamber was controlled by feeding the water molecules with the flow of N_2. All the devices were measured at RT in darkness. For the transfer curve measurements, V_{DS} of 10 V was employed. In this study, V_{th} is defined as the gate voltage of the normalized drain current ($I_{DS}/(W/L)$) reaching 100 nA.

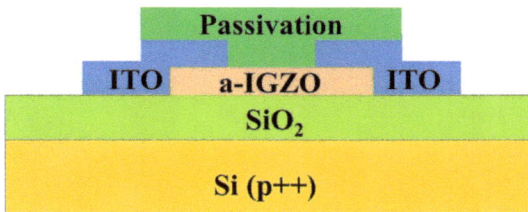

Figure 1. Schematic cross-section of the inverted staggered amorphous InGaZnO (a-IGZO) thin-film transistors (TFTs).

3. Results and Discussion

Figure 2a,c shows the time evolution of the transfer characteristics of the unpassivated a-IGZO TFTs under PGBS as well as the relative humidity (RH) of 10%, 50%, and 90%, respectively. During the PGBS tests, direct voltage of +20 V was applied to the gate electrodes for a period and then the transfer curves were instantly measured. With the increase in the stress time, the transfer curve positively shifted, which was apparently influenced by RH. In order to quantitatively describe the stable properties of a-IGZO TFTs under PGBS, we defined a useful term ΔV_{th}, the difference between the V_{th} under stress and its initial value. The ΔV_{th} values under various RH were extracted and listed in Figure 2d. After 4500 s of PGBS test, the positive V_{th} shifts of 5 V, 11.5 V, and 4.5 V were observed under RH = 10%, 50%, and 90%, respectively. It is worth noting that the largest ΔV_{th} occurred at RH = 50% (as shown in Figure 2d), which is consistent with our previous report [22].

The positive V_{th} shift of a-IGZO TFTs under PGBS was attributed to charge trapping at the dielectric/channel interface (front-channel effect) [3–6], ambient atmosphere effects at the back surface (back-channel effect) [7–14], or bulk trapping in the IGZO bulk (bulk effect) [15,16]. According to our previous work [22], The biggest V_{th} shift at RH = 50% is mainly attributed to the competition of oxygen (or moisture) adsorption/desorption at the IGZO back surface during PGBS tests. This result indicates that RH = 50% is the severest condition to characterize the bias-stress stability of a-IGZO TFTs.

In addition, we measured the negative gate bias-stress (NGBS) instability of a-IGZO TFTs at RH = 50%, as shown in Figure 3. During the NGBS tests, a direct voltage of −20 V was applied to the gate electrodes for a period and then the transfer curves were instantly measured. After 4500 s of NGBS test, nearly no V_{th} shift was observed. When a negative voltage was applied to the gate electrode of a-IGZO TFTs, the oxygen atoms in a-IGZO tended to be repelled into the ambience, leading to negative shifts of V_{th} [25]. However, this process might have been effectively prohibited by the moisture-assisted oxygen adsorption [11,12,22], especially when the ambient RH was high. Therefore, no evident V_{th} shifts were exhibited during the NGBS tests in this study.

In this study, we deliberately adopted the severest measurement condition (RH = 50%) to examine the influence of PV layers on the bias-stress stability of a-IGZO TFTs. Since the devices were rather stable during the NGBS tests, only PGBS stabilities were characterized for the following studies.

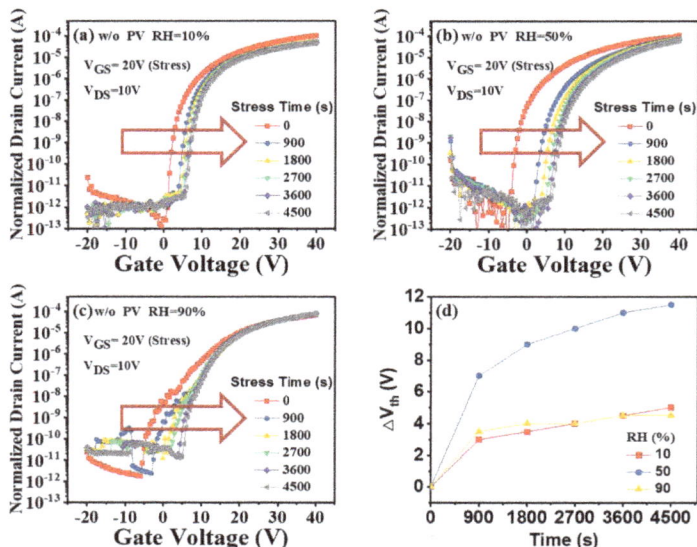

Figure 2. Transfer characteristics of the unpassivated a-IGZO TFTs as a function of the positive gate bias-stress (PGBS) time under relative humidity (RH) of (**a**) 10%, (**b**) 50%, and (**c**) 90%, respectively; (**d**) variations of the ΔV_{th} with PGBS time for the a-IGZO TFT devices.

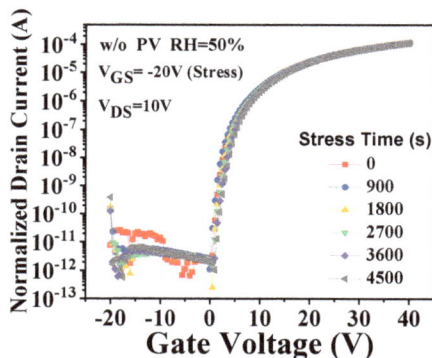

Figure 3. Transfer characteristics of the unpassivated a-IGZO TFTs as a function of the negative gate bias-stress (NGBS) time under RH = 50%.

It is well-known that PV layers can effectively improve the stability of TFT devices, whereas the exact physical mechanism involved is still not very clear. However, we may phenomenally describe the dependence of V_{th} shift (ΔV_{th}) during PGBS tests on PV layer thickness (d) as follows [19],

$$\Delta V_{th} = \alpha \cdot e_{\tau}^{d} + \beta \qquad (1)$$

where β is the V_{th} shift affected by charge trapping, bulk trapping, and plasma damage, α is a constant relating to the V_{th} shift affected by ambient atmosphere, and ε is the characteristic length related to the gas diffusion. When d is larger than ε, the ambient gases hardly influence the PGBS stability of a-IGZO TFTs. In other words, the characteristic length ε is the critical dimension for the ambient atmosphere effect during PGBS tests. From an application perspective, a small ε is usually preferred.

To further investigate the ambient effects during PGBS tests, the a-IGZO TFTs were applied using PV layers with different thicknesses. SiO_2, one of the most popular dielectric materials in TFT fabrications, was used to passivate the devices here. For comparison purposes, Al_2O_3, another dense material [23], was also adopted as PV layers for the a-IGZO TFTs in this study. The water vapor transmission rate (WVTR) and oxygen transmission rate (OTR) are reported to be inversely proportional to the PV layer thickness [24]. To analyze the influence of PV layers on PGBS stability of a-IGZO TFTs in depth, SiO_2 and Al_2O_3 films with different thicknesses (0–30 nm) were deposited to passivate the devices.

Figure 4a,e shows the PGBS time evolution of the transfer characteristics of the a-IGZO TFTs with a SiO_2 PV layer thickness of 0 nm, 5 nm, 10 nm, 20 nm, and 30 nm, respectively. We noticed that the passivated a-IGZO TFTs exhibited a similar tendency to that of the unpassivated device, i.e., with the increase in the stress time, the transfer curve gradually shifted in the positive direction. However, the a-IGZO TFTs with SiO_2 PV showed more stable properties during the PGBS tests. To describe this tendency more clearly, we extracted the V_{th} shifts and listed them in Figure 4f. When the PV layer thickness increased from 0 nm to 30 nm, the ΔV_{th} decreased evidently from 12 V to nearly 0.1 V after 4500 s of bias stress test. This can be attributed to the PV layer barrier effect, i.e., preventing the exchange of O_2/H_2O molecules between the channel layers and the ambient atmosphere. When the PV layer thickness was larger than 5 nm, the V_{th} of the a-IGZO TFTs barely changed. This can be understood by considering the concept of characteristic length (see (1)) in PV layers, which was about 5 nm here. When the SiO_2 PV layer thickness was smaller than the characteristic length, the O_2 molecules easily diffused from the atmosphere into a-IGZO (the H_2O diffused inversely) under PGBS, resulting in positive V_{th} shifts of the a-IGZO TFTs. As the PV layer thickness was larger than ε, the diffusion of O_2/H_2O molecules through the PV layers became rather difficult. This is why the device with a thicker PV layer showed less degradation of its electrical behavior. Since a sufficiently thick PV could nearly eliminate the V_{th} shifts (as shown in Figure 4f), we can assume that the ambient atmosphere effect, rather than charge trapping, dominated the instability of a-IGZO TFTs during the PGBS tests in this study.

For comparison purposes, we also measured the PGBS stability of the a-IGZO TFTs passivated by Al_2O_3 PV layers with a thickness of 0 nm, 5 nm, 10 nm, 20 nm, and 30 nm, respectively, as shown in Figure 5a–e. We may observe that a fairly similar tendency to the case of SiO_2-passivated devices was obtained here, i.e., the a-IGZO TFTs under PGBS became increasingly stable with the increase in the Al_2O_3 PV layer thickness. Meanwhile, for both Al_2O_3 and SiO_2 PVs, the transfer curve positively shifted as the PV layer thickness increased, which can be attributed to the extra interface states generated during the PV depositions [26,27]. However, the Al_2O_3 PV layer also exhibited something different. As shown in Figure 5e, for the device with a thick PV layer ($\geq$20 nm), its leakage current gradually rose with the increase in the stress time. This phenomenon was probably due to the plasma bombardment on the surface. Since the deposition rate of Al_2O_3 (0.7 nm/min) was smaller than that of SiO_2 (1.5 nm/min), more sputtering time was needed for the deposition of the Al_2O_3 PV layers, leading to more serious plasma damage at the back channels. What is more, the ion bombardment of the plasma can result in a positive V_{th} shift [17,18], which explains why the leakage current of the devices with thick Al_2O_3 PV layers increased during the PGBS tests.

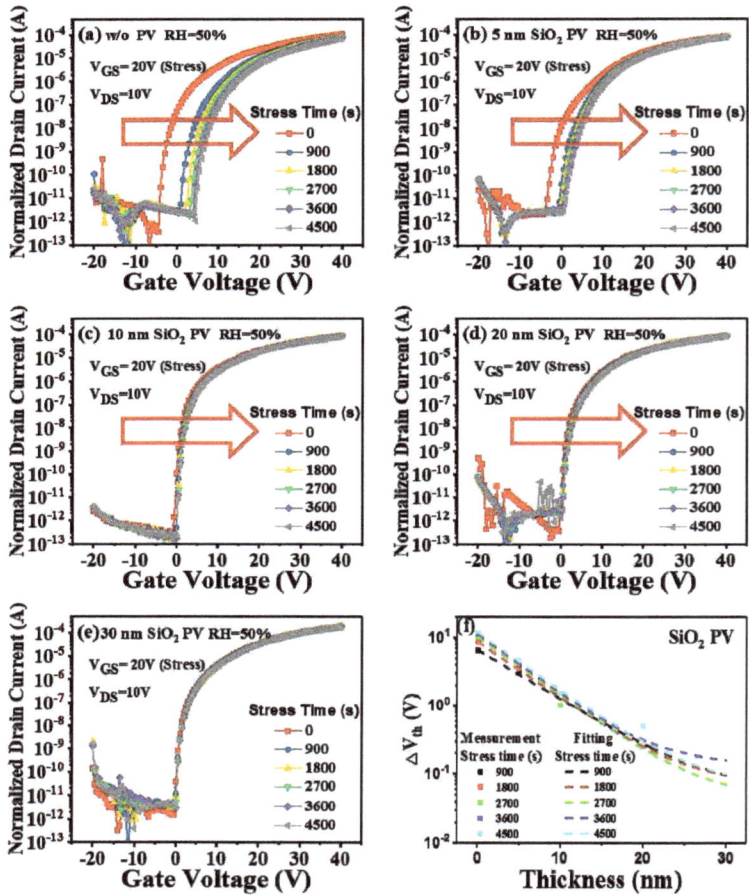

Figure 4. Stress-time dependence of the transfer characteristics of the a-IGZO TFTs with a SiO_2 PV layer thickness of (**a**) 0 nm, (**b**) 5 nm, (**c**) 10 nm, (**d**) 20 nm, and (**e**) 30 nm, respectively; (**f**) experimental data and fitting curves of the ΔV_{th} under PGBS as a function of PV layer thickness of the a-IGZO TFTs.

To precisely denote the influence of the Al_2O_3 PV layer on the PGBS stability of a-IGZO TFTs, the V_{th} shifts were extracted and listed in Figure 5f. Compared with the data shown in Figure 4f, we may note that the Al_2O_3 PV layers had an inferior barrier function to SiO_2. When the Al_2O_3 PV layer thickness was larger than 10 nm, the V_{th} shift of the devices changed slightly, indicating that the characteristic length of the Al_2O_3 PV layer was around 10 nm. When the PV layer thickness reached 30 nm, the ΔV_{th} became much smaller (~1 V), again confirming that the ambient atmosphere effect dominated during the PGBS tests in this study.

So far, we have obtained two important experimental results: (1) the PGBS stability of a-IGZO TFTs gradually improved with the increase in PV layer thickness; (2) the SiO_2 PV layer exhibited a better improvement effect on the PGBS stability than Al_2O_3. In order to discuss the theoretical origin of these results, we extracted the critical parameters in (1) of the PV layers. We fit the measurement data of SiO_2 PV and Al_2O_3 PV with (1), as shown in Figures 4 and 5, respectively. One may observe that the fitting curves agreed well with the measurement data, from which the fitting parameters were obtained and summarized in Table 1.

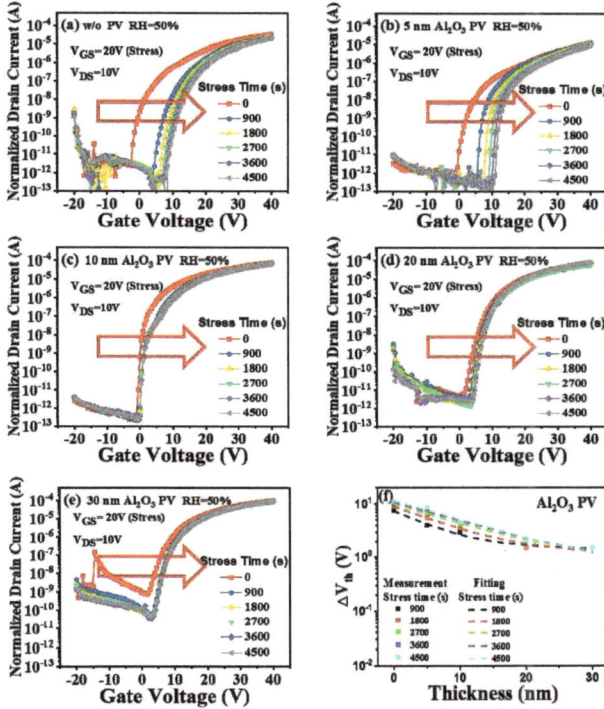

Figure 5. Stress-time dependence of the transfer characteristics of the a-IGZO TFTs with an Al_2O_3 PV layer thickness of (**a**) 0 nm, (**b**) 5 nm, (**c**) 10 nm, (**d**) 20 nm, and (**e**) 30 nm, respectively; (**f**) experimental data and fitting curves of the ΔV_{th} under PGBS as a function of the PV layer thickness of the a-IGZO TFTs.

Table 1. Fitting parameters of the PV layers used for a-IGZO TFTs.

Materials	Stress Time (s)	α (V)	β (V)	ε (nm)
SiO_2	900	6.47	0.06	5.93
	1800	8.46	0.07	5.22
	2700	10	0.04	5.03
	3600	10.86	0.14	4.86
	4500	11.45	0.06	5.13
Al_2O_3	900	6.07	1.39	9.24
	1800	7.95	1.09	9.53
	2700	9.31	0.84	10.21
	3600	10.46	0.30	11.23
	4500	11.07	0.25	11.06

As shown in Table 1, the α values of both PV layers were much larger than the β values for the same stress time, indicating that the ambient atmosphere effect instead of charge trapping dominated during the PGBS tests in this study. Therefore, with the increase in PV layer thickness, the ambient atmosphere effect was more strongly prevented, resulting in better PGBS stability of a-IGZO TFTs. The α value increased with the increase in the bias time, whereas the ε remained nearly unchanged. The increase in α resulted from more O_2/H_2O exchange between the device back channels and the ambience, leading to a larger V_{th} shift. Most importantly, SiO_2 and Al_2O_3 exhibited quite different characteristic length (ε) values, as shown in Table 1. The characteristic length of the SiO_2 PV layers (~5 nm) was far smaller than that of Al_2O_3 (~10 nm), leading to better improvement of the PGBS stability of a-IGZO TFTs by SiO_2 PV layers than Al_2O_3. Therefore, based on our results, the sputtered SiO_2, rather than the sputtered Al_2O_3, should be preferred to passivate a-IGZO TFTs in applications of FPDs.

4. Conclusions

The transfer curve of a-IGZO TFTs shifted positively during the PGBS tests, which could effectively be improved by applying PV layers. In this work, both SiO_2 and Al_2O_3 films with different thicknesses were used to passivate the a-IGZO TFTs, indicating that the ambient atmosphere effect rather than charge trapping dominated the V_{th} shifts during the PGBS tests. A simple model was used to theoretically discuss the related physical mechanism. With the increase in PV layer thickness, the devices became increasingly stable, as a result of the stronger prevention of the ambient atmosphere effect. When the PV layer thickness reached the characteristic length, the variation in V_{th} became quite small. The SiO_2 PV layer showed a better improvement effect than the Al_2O_3 PV layer because the former had a smaller characteristic length.

Author Contributions: Y.Z. fabricated and measured all the a-IGZO TFTs. Y.Z. and C.D. designed the experiments and contributed to the theoretical explanations. The manuscript was written by Y.Z. and C.D.

Funding: This work was supported by the Natural Science Foundation of China (grant No. 61474075).

Conflicts of Interest: The authors declare no conflict of interest.

References

1. Nomura, K.; Ohta, H.; Takagi, A.; Kamiya, T.; Hirano, M.; Hosono, H. Room-temperature fabrication of transparent flexible thin-film transistors using amorphous oxide semiconductors. *Nature* **2004**, *432*, 488–492. [CrossRef] [PubMed]
2. Kim, M.K.; Jeong, J.H.; Lee, H.J.; Ahn, T.K.; Shin, H.S.; Park, J.S.; Jeong, J.K.; Mo, Y.G.; Kim, H.D. High mobility bottom gate InGaZnO thin film transistors with SiOx etch stopper. *Appl. Phys. Lett.* **2007**, *90*, 212114. [CrossRef]
3. Suresh, A.; Muth, J.F. Bias stress stability of indium gallium zinc oxide channel based transparent thin film transistors. *Appl. Phys. Lett.* **2008**, *92*, 033502. [CrossRef]
4. Lee, J.M.; Cho, I.T.; Lee, J.H.; Kwon, H.I. Bias-stress-induced stretched-exponential time dependence of threshold voltage shift in InGaZnO thin film transistors. *Appl. Phys. Lett.* **2008**, *93*, 093504. [CrossRef]
5. Chen, T.C.; Chang, T.C.; Hsieh, T.Y.; Lu, W.S.; Jian, F.Y.; Tsai, C.T.; Huang, S.Y.; Lin, C.S. Investigating the degradation behavior caused by charge trapping effect under DC and AC gate-bias stress for InGaZnO thin film transistor. *Appl. Phys. Lett.* **2011**, *99*, 022104. [CrossRef]
6. Kim, S.J.; Lee, S.Y.; Lee, Y.W.; Kuk, S.H.; Kwon, J.Y.; Han, M.K. Effect of Charge Trapping/Detrapping on Threshold Voltage Shift of IGZO TFTs under AC Bias Stress. *Electrochem. Solid-State Lett.* **2012**, *15*, H108–H110. [CrossRef]
7. Sung, S.Y.; Choi, J.H.; Han, U.B.; Lee, K.C.; Lee, J.H.; Kim, J.J.; Lim, W.; Pearton, S.J.; Norton, D.P.; Heo, Y.W. Effects of ambient atmosphere on the transfer characteristics and gate-bias stress stability of amorphous indium-gallium-zinc oxide thin-film transistors. *Appl. Phys. Lett.* **2010**, *96*, 102107. [CrossRef]
8. Jeong, J.K.; Yang, H.W.; Jeong, J.H.; Mo, Y.G.; Kim, H.D. Origin of threshold voltage instability in indium-gallium-zinc oxide thin film transistors. *Appl. Phys. Lett.* **2008**, *93*, 123508. [CrossRef]
9. Liu, P.T.; Chou, Y.T.; Teng, L.F. Environment-dependent metastability of passivation-free indium zinc oxide thin film transistor after gate bias stress. *Appl. Phys. Lett.* **2009**, *95*, 233504. [CrossRef]
10. Chen, Y.C.; Chang, T.C.; Li, H.W.; Chen, S.C.; Lu, J.; Chung, W.F.; Tai, Y.H.; Tseng, T.Y. Bias-induced oxygen adsorption in zinc tin oxide thin film transistors under dynamic stress. *Appl. Phys. Lett.* **2010**, *96*, 262104. [CrossRef]

11. Zhang, J.; Li, X.F.; Lu, J.G.; Zhou, N.J.; Guo, P.J.; Lu, B.; Pan, X.H.; Chen, L.X.; Ye, Z.Z. Water assisted oxygen absorption on the instability of amorphous InAlZnO thin-film transistors. *RSC Adv.* **2014**, *4*, 3145–3148. [CrossRef]

12. Chung, W.F.; Chang, T.C.; Li, H.W.; Chen, S.C.; Chen, Y.C.; Tseng, T.Y.; Tai, Y.H. H_2O-Assisted O_2 Adsorption in Sol-Gel Derived Amorphous Indium Gallium Zinc Oxide Thin Film Transistors. *Electrochem. Solid State Lett.* **2011**, *14*, H235–H237. [CrossRef]

13. Huang, S.Y.; Chang, T.C.; Chen, M.C.; Chen, S.C.; Tsai, C.T.; Hung, M.C.; Tu, C.H.; Chen, C.H.; Chang, J.J.; Liau, W.L. Effects of Ambient Atmosphere on Electrical Characteristics of Al_2O_3 Passivated InGaZnO Thin Film Transistors during Positive-Bias-Temperature-Stress Operation. *Electrochem. Solid State Lett.* **2011**, *14*, H177–H179. [CrossRef]

14. Chen, F.H.; Pan, T.M.; Chen, C.H.; Liu, J.H.; Lin, W.H.; Chen, P.H. Two-step Electrical Degradation Behavior in α-InGaZnO Thin-film Transistor Under Gate-bias Stress. *IEEE Electron Device Lett.* **2013**, *34*, 635–637. [CrossRef]

15. Cho, E.N.; Kang, J.H.; Yun, I. Effects of channel thickness variation on bias stress instability of InGaZnO thin-film transistors. *Microelectron. Reliab.* **2011**, *51*, 1792–1795. [CrossRef]

16. Li, J.; Ding, X.W.; Zhang, J.H.; Zhang, H.; Jiang, X.Y.; Zhang, Z.L. Improving electrical performance and bias stability of HfInZnO-TFT with optimizing the channel thickness. *AIP Adv.* **2013**, *3*, 102132. [CrossRef]

17. Lin, W.K.; Liu, K.C.; Chen, J.N.; Hu, S.C.; Chang, S.T. The influence of fabrication process on top-gate thin-film transistors. *Thin Solid Films* **2011**, *519*, 5126–5130. [CrossRef]

18. Lin, W.K.; Liu, K.C.; Chang, S.T.; Li, C.S. Room temperature fabricated transparent amorphous indium zinc oxide based thin film transistor using high-κ HfO_2 as gate insulator. *Thin Solid Films* **2011**, *520*, 3079–3083. [CrossRef]

19. Li, G.T.; Yang, B.R.; Liu, C.; Lee, C.Y.; Tseng, C.Y.; Lo, C.C.; Lien, A.; Deng, S.Z.; Shieh, H.P.D.; Xu, N.S. Positive gate bias instability alleviated by self-passivation effect in amorphous InGaZnO thin-film transistors. *J. Phys. D Appl. Phys.* **2015**, *48*, 475107. [CrossRef]

20. Fan, C.L.; Shang, M.C.; Li, B.J.; Lin, Y.Z.; Wang, S.J.; Lee, W.D.; Hung, B.R. Teflon/SiO_2 bilayer passivation for improving the electrical reliability of oxide TFTs fabricated using a new two-photomask self-alignment process. *Materials* **2015**, *8*, 1704–1713. [CrossRef] [PubMed]

21. Li, J.; Zhou, F.; Lin, H.P.; Zhu, W.Q.; Zhang, J.H.; Jiang, X.Y.; Zhang, Z.L. Effect of reactive sputtered SiO_x passivation layer on the stability of InGaZnO thin film transistors. *Vacuum* **2012**, *86*, 1840–1843. [CrossRef]

22. Zhou, Y.; Xu, J.; Xie, H.; Zhang, L.; Liu, G.; Tong, X.; Dong, C. Moisture Dominant Electrical Degradation of Amorphous InGaZnO Thin Film Transistors under Positive Bias Stress. In Proceedings of the International Display Workshop (IDW)'17, Sendai, Japan, 6–8 December 2017; pp. 424–426.

23. Wu, J.; Chen, Y.; Zhou, D.; Hu, Z.; Xie, H.; Dong, C. Sputtered oxides used for passivation layers of amorphous InGaZnO thin film transistors. *Mater. Sci. Semicond. Process.* **2014**, *29*, 277–282. [CrossRef]

24. Graff, G.L.; Burrows, P.E.; Williford, R.E.; Praino, R.F. Barrier Layer Technology for Flexible Displays. In *Flexible Flat Panel Display*, 1st ed.; Crawford, G.P., Ed.; Brown University: Providence, RI, USA, 2005; pp. 57–60. ISBN 0-470-87048-6.

25. Chen, Y.C.; Chang, T.C.; Li, H.W.; Chen, S.C.; Chung, W.F.; Chen, Y.H.; Tai, Y.H.; Tseng, T.Y.; Yeh, F.S. Surface states related the bias stability of amorphous In–Ga–Zn–O thin film transistors under different ambient gasses. *Thin Solid Films* **2011**, *520*, 1432–1436. [CrossRef]

26. Chen, T.C.; Chang, T.C.; Hsieh, T.Y.; Tsai, C.T.; Chen, S.C.; Lin, C.S.; Hung, M.C.; Tu, C.H.; Chang, J.J.; Chen, P.L. Light-induced instability of an InGaZnO thin film transistor with and without SiO_x passivation layer formed by plasma-enhanced-chemical-vapor-deposition. *Appl. Phys. Lett.* **2010**, *97*, 192103. [CrossRef]

27. Ohara, H.; Sasaki, T.; Noda, K.; Ito, S.; Sasaki, M.; Endo, Y.; Yoshitomi, S.; Sakata, J.; Serikawa, T.; Yamazaki, S. 4.0-inch active-matrix organic light-emitting diode display integrated with driver circuits using amorphous In–Ga–Zn-Oxide thin-film transistors with suppressed variation. *Jpn. J. Appl. Phys.* **2010**, *49*, 03CD02. [CrossRef]

micromachines

MDPI

Article

Step-Double-Zone-JTE for SiC Devices with Increased Tolerance to JTE Dose and Surface Charges

Yifei Huang, Ying Wang *, Xiaofei Kuang, Wenju Wang, Jianxiang Tang and Youlei Sun

Key Laboratory of RF Circuits and Systems, Ministry of Education, Hangzhou Dianzi University, Hangzhou 310018, China; 162040160@hdu.edu.cn (Y.H.); kuangxiaofei@hdu.edu.cn (X.K.); 162040172@hdu.edu.cn (W.W.); hztangjx@hdu.edu.cn (J.T.); 161040013@hdu.edu.cn (Y.S.)
* Correspondence: wangying7711@yahoo.com; Tel.: +86-0571-8691-9135

Received: 10 October 2018; Accepted: 20 November 2018; Published: 22 November 2018

check for updates

Abstract: In this paper, an edge termination structure, referred to as step-double-zone junction termination extension (Step-DZ-JTE), is proposed. Step-DZ-JTE further improves the distribution of the electric field (EF) by its own step shape. Step-DZ-JTE and other termination structures are investigated for comparison using numerical simulations. Step-DZ-JTE greatly reduces the sensitivity of breakdown voltage (BV) and surface charges (SC). For a 30-μm thick epi-layer, the optimized Step-DZ-JTE shows 90% of the theoretical BV with a wide tolerance of 12.2×10^{12} cm^{-2} to the JTE dose and 85% of the theoretical BV with an improved tolerance of 3.7×10^{12} cm^{-2} to the positive SC are obtained. Furthermore, when combined with the field plate technique, the performance of the Step-DZ-JTE is further improved.

Keywords: edge termination; silicon carbide (SiC); junction termination extension (JTE); breakdown voltage (BV)

---

1. Introduction

Silicon carbide (SiC), a representative of the third generation of semiconductor materials, is a promising candidate for power devices due to its superior properties such as wide bandgap, high breakdown electric field, high thermal conductivity, and high drift saturation speed [1–5]. Hence, SiC devices are more suitable than silicon counterparts for high-voltage, high-frequency, and high-temperature applications. However, the potential performance (e.g., high blocking) of SiC materials is limited due to the presence of the effect of field crowding at the device edge.

To achieve high breakdown voltages (BV) for the planar junction close to its theoretical voltage, using a proper edge termination structures is essential. In the past few decades, a large number of edge termination structures have been investigated and applied in SiC power devices, including field plate (FP) [6,7], floating field rings (FFRs) [8–10], and junction termination extension (JTE) [11–13]. FP is easy to design and fabricate, but introduces electric field (EF) peaks at the end, which limits their application to high-voltage devices. FFRs are widely used in power devices because they can be formed at the same time as the P+ region in a PIN or junction barrier Schottky diode, while this reduces the number of processing steps, but FFRs are more demanding to design with specific ring requirements such as width, spacing, and number of rings. Among them, JTE is a prevalent and highly efficient edge termination structure for SiC power devices. The effective JTE can sustain high BV, but is strongly dependent on precise JTE dose control, which is a big challenge. The BV for conventional single-zone junction termination extension (SZ-JTE) is very sensitive to JTE dose. Feng [14] investigated PIN diodes in 4H-SiC with different terminal structures. The results showed that, for 90% of the theoretical BV, a conventional SZ-JTE obtained a narrow JTE dose tolerance of 1.0×10^{12} cm^{-2}. Therefore, many modified forms of JTE have been proposed to improve the sensitivity, such as guard ring-assisted

JTE (GA-JTE) [15–17], double-zone JTE (DZ-JTE) [18,19], multiple-zone JTE (MZ-JTE) [20,21], etched JTE [22,23], counter-doped JTE (CD-JTE) [24], and mesa combined with JTEs [25,26]. Feng [14] also reported that the JTE dose tolerance (4.8×10^{12} cm^{-2}) in the conventional DZ-JTE was improved compared with the conventional SZ-JTE. Huang [24] proposed and investigated PIN diodes with CD-JTE and other conventional terminal structures. The simulation results in his paper showed the CD-JTE with a JTE dose tolerance of 11.0×10^{12} cm^{-2} was greatly improved compared with other structures.

In this paper, an edge termination structure called Step-DZ-JTE for 4H-SiC PIN diode is proposed and investigated. Combined with conventional DZ-JTE, the proposed structure changes the shape of DZ-JTE to a step type to adjust the distribution of the electric field. On the one hand, the Step-DZ-JTE does not add to the number of p-type implants. On the other hand, the simulation results show that a wide tolerance of 12.2×10^{12} cm^{-2} is obtained for the Step-DZ-JTE, which is greatly improved over the conventional DZ-JTE and slightly improved compared with the CD-JTE. Moreover, when combined with the FP technology, the performance of the Step-DZ-JTE is further improved, as verified by numerical simulations.

2. Materials and Methods

2.1. Device Structure

In this section, a 4H-SiC PIN diode with a 30-μm epi-layer doped at 3×10^{15} cm^{-3} can attain 4000 V from the ideal parallel junction [27]. All termination structures for PIN diodes were investigated using Silvaco TCAD 2-D device simulations (atlas 5.22.1.R, Silvaco Inc., Santa Clara, CA, USA). The breakdown condition was defined as the point at which the calculated impact ionization integral reaches unity [23]. The major models used in simulations include Schokley-Read-Hall (SRH), Auger recombination, impact ionization, and incomplete ionization. Cross-sectional views of PIN diodes with three termination structures are shown in Figure 1, including: (a) single-zone JTE (SZ-JTE); (b) double-zone JTE (DZ-JTE); (c) Step-DZ-JTE; (d) Step-DZ-JTE with FP. The feature of Step-DZ-JTE is a step structure, which improves EF distribution by adjusting the depth and length of the step JTE. In addition, Step-DZ-JTE with FP can reduce the amount of EF crowding near the main junction in a low JTE dose by adding an anode FP. In this figure "D_s", "L_s", "W_{fp}" are the depth and length of step JTE1, and the length of the anode FP, respectively. Among these structures, the length of termination region is fixed at 100 μm, where the simulated BV of the JTE starts to saturate. The values of the major optimized parameters of the proposed structure are summarized in Table 1.

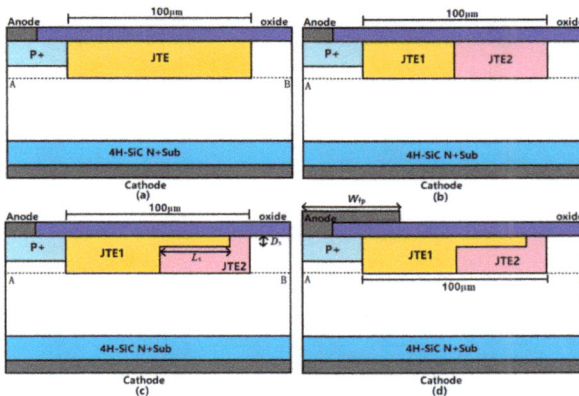

Figure 1. Schematic structures of (**a**) single-zone junction termination extension (SZ-JTE); (**b**) double-zone JTE (DZ-JTE); (**c**) Step-DZ-JTE; (**d**) Step-DZ-JTE with field plate (FP). The n- epi-layer is 30-μm thick and 2×10^{15} cm^{-3} doped.

Table 1. Major optimized parameters of the proposed structure.

Parameter	Value
P+ anode junction depth	0.6 μm
Junction termination extension (JTE) junction depth	0.8 μm
Depth of the step JTE (D_s)	0.3 μm
Length of the step JTE (L_s)	45 μm
Length of the anode FP (W_{fp})	20 μm
Thickness of drift	30 μm
n- drift dopant concentration (N_D)	3.0×10^{15} cm^{-3}
p$^+$ anode dopant concentration (N_A)	1.0×10^{19} cm^{-3}

2.2. Fabrication Procedure

Figure 2 shows a feasible fabrication procedure for building a step JTE in the Step-DZ-JTE. As shown in Figure 2a, an n- 4H-SiC epitaxial layer is first grown on an n+ 4H-SiC substrate. Then, SiO$_2$ layer as the mask materials are thermally grown on the epitaxial layer and a photoresist is patterned on the mask to form a JTE2 window [28]. Next, multiple aluminum implantations are applied to form the JTE2 region. Then, a graphite cap layer is grown on the surface after removing the mask material to prevent the sublimation and roughening of the surface during the next annealing [28,29]. Next the implantations of the JTE2 are activated by high temperature annealing to activate acceptor impurities and form a box profile, as shown in Figure 2b. In order to form the step distribution of the JTE1, the step mask, as shown in Figure 2c is the key to the formation of the JTE and its shape can be form by the etching process. Similarly, the JTE1 region is formed using ion implantation through the mask and then activation annealing. Finally, the main P+ region is formed using ion implantation with high doses, as shown in Figure 2d. All annealing conditions are implemented at the temperature of 1650 °C under argon ambient with the graphite cap [30]. This process is easier to implement than etching, and it avoids the extra interface charges caused by filling the dielectric after etching.

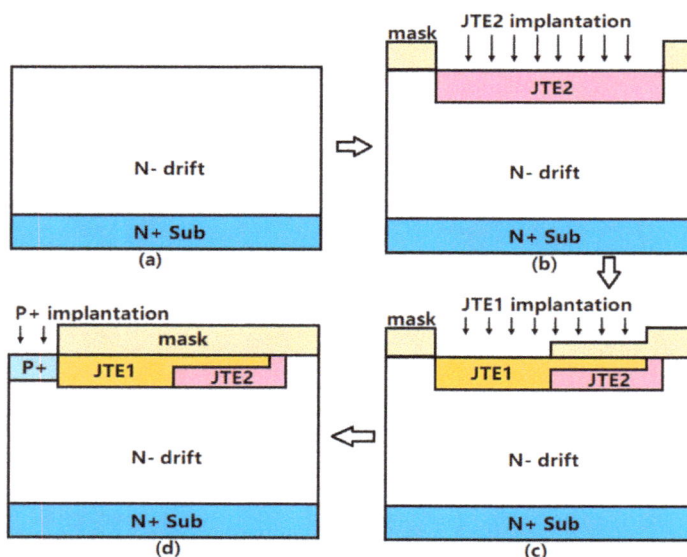

Figure 2. Fabrication procedure of making step JTE in the Step-DZ-JTE. (**a**) Base layers. (**b**) JTE2 region formed by ion implantation. (**c**) JTE1 region fromed by ion implantation with the specific mask. (**d**) Main junction P+ formation.

3. Results

3.1. Simulation Optimization of the DZ-JTE

Based on the optimized conventional SZ-JTE, this section discusses the optimization of the DZ-JTE, which mainly involves the length and dose of the double JTE. Figure 3 shows the simulated BV versus the ratio of dose1 in JTE1 to dose2 in JTE2 for DZ-JTE. In the higher JTE1 dose range, the ratio of doses has a significant effect on the BV. The BV increases as the ratio increases, and then drops sharply. There is a reasonable ratio value of doses to avoid a sharp decrease in the BV. As can be seen from the figure, the optimized ratio value of dose1/dose2 is 3.

Based on the optimized ratio above, the simulated BV versus JTE1 length for DZ-JTE with different JTE1 dose is shown in Figure 4. We see that JTE1 length has little effect on BV in the lower of JTE1 dose range. This phenomenon is explained by the fact that too low a concentration makes no difference between JTE1 and JTE2. In other JTE1 doses, as JTE1 length increases, the curve gradually rises. When JTE1 length is 50 μm, the curve reaches its highest point. In addition, the curve drops rapidly as JTE1 length exceed 50 μm. This is because, when the JTE1 length is too long, the JTE2 length is shorter and the effect of JTE2 (alleviating the EF of JTE1) is reduced.

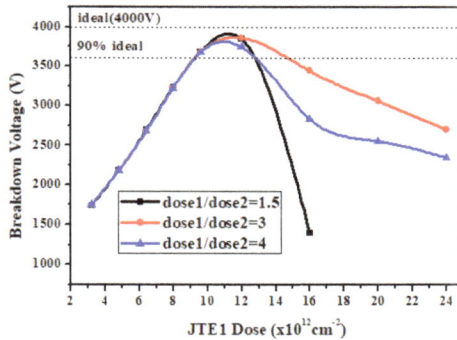

Figure 3. Simulated breakdown voltage (BV) versus ratio of the doses with different doses for the DZ-JTE optimization.

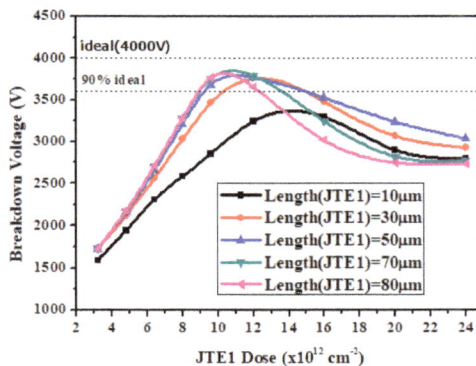

Figure 4. Simulated BV versus JTE1 length with different doses for the DZ-JTE optimization.

3.2. Simulation Optimization of the Step-DZ-JTE with FP

Based on the optimized DZ-JTE, this section optimizes the length and depth of the step JTE1 for the proposed Step-DZ-JTE. Figure 5 shows the simulated BV versus the depth (D_s) and the length (L_s)

of step JTE1 for Step-DZ-JTE with FP. As can be seen from Figure 5a, the curve of the BV versus JTE1 dose rises first and then declines as the D_s increases in higher of JTE1 dose range (i.e., >12 × 10^{12} cm^{-2}). This is because the EF is more concentrated at the end of the step JTE1 with the D_s increases, leading to premature breakdown. When the D_s is 0.3 μm, the Step-DZ-JTE with FP attains the maximum value of the BV, meaning the optimized value of D_s is 0.3 μm. Figure 5b shows the simulated BV versus the length (L_s) of step JTE1 for Step-DZ-JTE. When JTE1 dose is lower than 12 × 10^{12} cm^{-2}, the L_s has no effect on the relationship between the BV and JTE dose. In the other range of JTE1 dose, the value of 45 μm is a critical value of the L_s. Regardless of whether the L_s is larger or small than the value, the curves of the BV versus JTE1 dose are lower than the curve corresponding to 45 μm. It can be seen from the figure that the longer the L_s is, the faster the curves fall. This phenomenon can be explained by the fact that the longer the L_s is, the higher the carrier concentration is in the step JTE1, resulting in EF crowding at the end of the step JTE1 in the reverse blocking state.

Figure 6 shows the effect of FP on the simulated BV for the Step-DZ-JTE with FP. The maximum BV is obtained when the FP length (W_{fp}) is 20 μm. The insets of Figure 6 show simulated EF distribution at breakdown with different W_{fp}. As shown in inset (a), the peak EF occurred near the main junction when the W_{fp} is less than 20 μm (e.g., 5 μm). As the W_{fp} increases, the EF crowding near the main can be effectively suppressed by the FP as shown in inset (b). The oxide field is 2.77 MV/cm, shown in inset (b), which is less than the oxide critical field (6 MV/cm in [31]). This means that there is no oxide degradation at breakdown. When the W_{fp} is more than the optimal length, the location of the peak EF shifts into the periphery of the FP, as shown in inset (c). As can be seen from inset (c), the oxide field (2.92 MV/cm) is also less than 6 MV/cm.

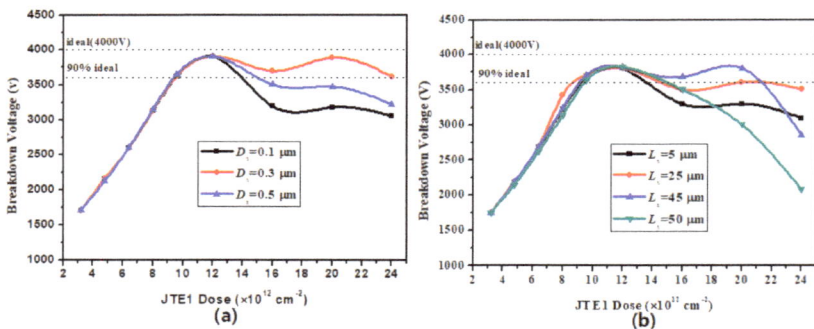

Figure 5. Simulated BV versus (**a**) the depth (D_s) and (**b**) the length (L_s) of the Step JTE for Step-DZ-JTE optimization.

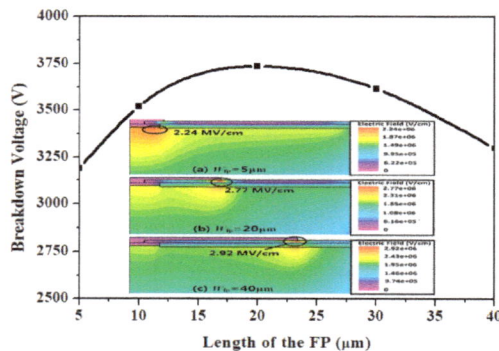

Figure 6. Simulated BV versus FP length (W_{fp}) for the Step-DZ-JTE with FP optimization.

3.3. Compare Electric Field Distribution

The distribution of the electric field under the reverse blocking characteristic can reflect the problem of the conventional terminal structures and the proposed Step-DZ-JTE can be clearly compare with them. Figure 7 compares the simulated EF distribution along the cutline of AB (shown in Figure 1) and on surface of the JTE for SZ-JTE, DZ-JTE, and Step-DZ-JTE with FP at the reverse blocking voltage of 3500 V when JTE1 dose is 2×10^{13} cm^{-2}. Regardless of the distribution of the EF along the bottom of the JTE (Figure 7a) or the distribution of the EF along the JTE surface (Figure 7b), the trend of their curves is generally the same. At this high JTE dose, the peak EF occurred at the end of JTE for the SZ-JTE, resulting in breakdown prematurely. The DZ-JTE can improve the EF at the terminal edge, but a new peak EF appears at the end of the JTE1. Under the effect of step JTE, the Step-DZ-JTE with FP further suppresses the EF crowding at the end of the JTE1. However, the effect of the FP for the Step-DZ-JTE with FP is very small. This is because that concentration of the acceptor carrier is too large at this high JTE dose, so that the effect of the FP absorbing part of the EF is not obvious.

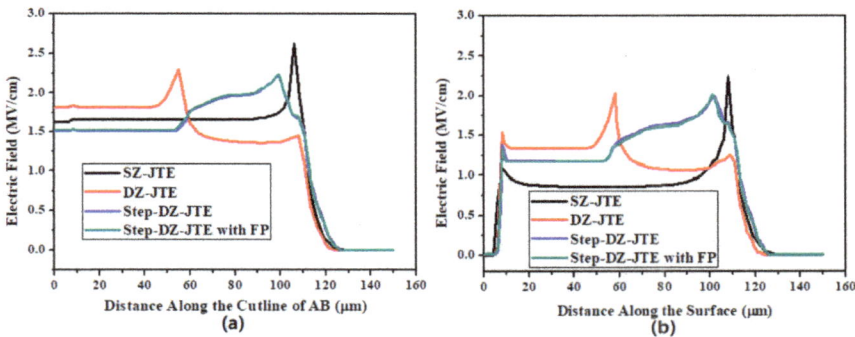

Figure 7. Simulated electric field distribution at OFF-state breakdown (**a**) along the cutline of AB and (**b**) on the surface for SZ-JTE, DZ-JTE, Step-DZ-JTE with FP (dose1/dose2 = 3, and JTE1 dose = 2×10^{13} cm^{-2}).

4. Discussion

In order to evaluate the performance of the Step-DZ-JTE, it is compared with other termination structures while examining the effects of JTE1 dose and surface charges (SC). First, we discuss the effect JTE dose has on the simulated BV. Figure 8 shows the BV as a function of the JTE1 dose and total JTE dose. As can be seen from Figure 8a, the SZ-JTE shows a very narrow JTE dose tolerance of 0.4×10^{12} cm^{-2} at 90% of the ideal BV. For the SZ-JTE, the percentage of positive and negative variation allowed to deviate from the optimized dose are +2.2% and −2.2% (positive means tolerance to allow for exceeding the optimized JTE dose; negative means tolerance to allow for less than the optimized JTE dose). By adopting two zones with different doses, the DZ-JTE shows a significant improvement of 4.1×10^{12} cm^{-2} at 90% of the ideal BV, of which the percentage of positive and negative variation are +17.5% and −17.0% The Step-DZ-JTE, proposed in this paper, is superior to the two termination structures mentioned above. The Step-DZ-JTE has a BV with reduced sensitivity to JTE1 dose and exhibits a wider JTE1 dose tolerance of 12.2×10^{12} cm^{-2} at 90% of the ideal BV of which the percentage variations are +75% and +18.4%. In the Step-DZ-JTE with FP, the curve coincides with the Step-DZ-JTE at the higher JTE dose. However, at a low JTE1 dose, the Step-DZ-JTE with FP performs better than the Step-DZ-JTE. This is because FP can relieve the EF in the main junction at low JTE1 dose. Thus, the Step-DZ-JTE creates a wide range of JTE1 dose at 90% of the ideal BV with 13.8×10^{12} cm^{-2}, of which the percentage variations are +75% and −35%. In addition, the total JTE dose can be calculated based on the previously optimized ratio value of dose1/dose2. The relationship between the BV and total dose is shown in Figure 8b. Total JTE dose tolerance for 90% of the ideal BV with four terminal structures are listed in Table 2.

On the other hand, the effect of SC on the BV is investigated since SC will affect the charge distribution and, thus, the EF distribution. In particular, positive SC has a large influence because the positive charges cancel the depleted acceptors in the JTE region. As a result, as shown in Figure 9, the SZ-JTE exhibits a very small tolerance to positive SC and obtains a positive charge density of 0.5×10^{12} cm^{-2} at 85% of the ideal BV. The DZ-JTE and Step-DZ-JTE show a BV with reduced sensitivity to SC and they obtain a positive charge density of 3.2×10^{12} cm^{-2} and 3.7×10^{12} cm^{-2} at 85% of the ideal BV, respectively. However, after adding the FP, the Step-DZ-JTE with FP shows the widest positive charge density of 5.5×10^{12} cm^{-2}. Table 2 summarizes the basic performance of the four different termination structures. Compared with DZ-JTE, the number of *p*-type ion implantations for the proposed Step-DZ-JTE did not increase, and the performance of the Step-DZ-JTE is greatly improved. Combined with FP technology, the Step-DZ-JTE with FP further reduces the sensitivity of BV to JTE doses and surface charges.

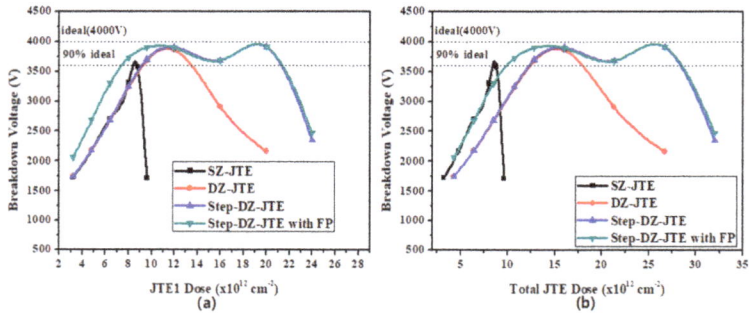

Figure 8. Simulated BV versus (**a**) JTE1 dose and (**b**) total JTE dose for SZ-JTE, DZ-JTE, Step-DZ-JTE without FP, and Step-DZ-JTE with FP.

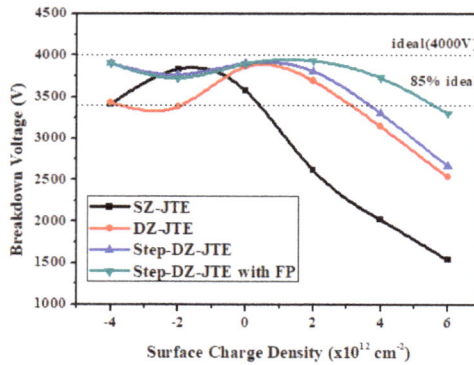

Figure 9. Simulated BV versus surface charges for SZ-JTE, DZ-JTE, Step-DZ-JTE without FP, and Step-DZ-JTE with FP.

Table 2. Basic properties of different JTEs.

Structures	SZ-JTE	DZ-JTE	Step-DZ-JTE	Step-DZ-JTE with FP
JTE total length (μm)	100	100	100	100
Number of p-type implant	1	2	2	2
JTE1 dose tolerance for 90% BV ($\times 10^{12}$ cm^{-2})	0.4	4.1	12.2	13.8
Total JTE dose tolerance for 90% BV (($\times 10^{12}$ cm^{-2})	0.4	5.6	16.3	18.4
The percentage of positive and negative variation	(+2.2%, −2.2%)	(+17.5%, −17.0%)	(+75%, −18.4%)	(+75%, −35%)
Max. positive SC density for 85% BV ($\times 10^{12}$ cm^{-2})	0.5	3.2	3.7	5.5

5. Conclusions

A Step-DZ-JTE edge termination scheme is proposed in this paper and the device simulation results show that it has superior terminal performance. A comparison of the simulation results for the SZ-JTE and DZ-JTE shows that the Step-DZ-JTE greatly reduces the sensitivity of the JTE dose and SC. For a 30-μm epi-layer, the Step-DZ-JTE can exhibits a wider JTE dose tolerance of 12.2×10^{12} cm^{-2} at 90% of the ideal BV and a positive charge density of 3.2×10^{12} cm^{-2} at 85% of the ideal BV. The Step-DZ-JTE with FP is introduced on the basis of a Step-DZ-JTE by adding an anode FP. The Step-DZ-JTE with FP further improves the effects of JTE dose and SC. Moreover, the Step-DZ-JTE with FP does not require an additional fabrication process. These performance improvements show that the Step-DZ-JTE with FP is a promising edge termination technique for SiC devices.

Author Contributions: Investigation—Y.H., W.W., J.T. and Y.S.; Conceptualization—Y.W.; Software—X.K.

Funding: This work was supported by the National Natural Science Foundation of China (No. 61774052) and the Excellent Youth Foundation of Zhejiang Province of China (No. LR17F040001).

Conflicts of Interest: The authors declare no conflict of interest.

References

1. Shenai, K.; Scott, R.S.; Baliga, B.J. Optimum semiconductors for high-power electronics. *IEEE Trans. Electron. Devices* **1989**, *36*, 1811–1823. [CrossRef]
2. Elasser, A.; Chow, T.P. Silicon carbide benefits and advantages for power electronics circuits and systems. *Proc. IEEE* **2002**, *90*, 969–986. [CrossRef]
3. Kimoto, T. Ultrahigh-voltage SiC devices for future power infrastructure. In Proceedings of the 2013 Proceedings of the European Solid-State Device Research Conference, Bucharest, Romania, 16–20 September 2013; pp. 22–29. [CrossRef]
4. Casady, J.B.; Johnson, R.W. Status of silicon carbide (SiC) as a wide-bandgap semiconductor for high-temperature applications: A review. *Solid-State Electron.* **1996**, *39*, 1409–1422. [CrossRef]
5. Buttay, C.; Raynaud, C.; Morel, H.; Civrac, G.; Locatelli, M.L.; Morel, F. Thermal stability of silicon carbide power diodes. *IEEE Trans. Electron. Devices* **2012**, *59*, 761–769. [CrossRef]
6. Shankar, B.; Gupta, S.K.; Taube, W.R.; Akhtar, J. High-k dielectrics based field plate edge termination engineering in 4H-SiC Schottky diode. *Int. J. Electron.* **2016**, *103*, 2064–2074. [CrossRef]
7. Tarplee, M.C.; Madangarli, V.P.; Zhang, Q.; Sudarshan, T.S. Design rules for field plate edge termination in SiC Schottky diodes. *IEEE Trans. Electron. Devices* **2001**, *48*, 2659–2664. [CrossRef]
8. Brunt, E.V.; Cheng, L.; O'Loughlin, M.; Capell, C. 22 kV, 1 cm^2, 4H-SiC n-IGBTs with improved conductivity modulation. In Proceedings of the International Symposium on Power Semiconductor Devices & Ic's, Waikoloa, HI, USA, 15–19 June 2014; pp. 358–361. [CrossRef]
9. Villamor-Baliarda, A.; Vanmeerbeek, P.; Roig, J.; Moens, P. Electric field unbalance for robust floating ring termination. *Microelectron. Reliab.* **2011**, *51*, 1959–1963. [CrossRef]
10. Sung, W.; Baliga, B.J. A Near Ideal Edge Termination Technique for 4500V 4H-SiC Devices: The Hybrid Junction Termination Extension (Hybrid-JTE). *IEEE Electron. Device Lett.* **2016**, *37*, 1609–1612. [CrossRef]
11. Mahajan, A.; Skromme, B.J. Design and optimization of junction termination extension (JTE) for 4H SiC high voltage Schottky diodes. *Solid State Electron.* **2005**, *49*, 945–955. [CrossRef]
12. Okayama, T.; Arthur, S.D.; Rao, R.R.; Kishore, K.; Rao, M.V. Stability and 2-D Simulation Studies of Avalanche Breakdown in 4H-SiC DMOSFETS with JTE. *IEEE Trans. Electron Devices.* **2008**, *55*, 489–494. [CrossRef]

13. Zhang, F.; Xing-Hua, F.U. Simulation study on 4H-SiC Schottky diode with JTE structure. In Proceedings of the 2014 16th European Conference on Power Electronics and Applications, Lappeenranta, Finland, 26–28 August 2012.

14. Feng, G.; Suda, J.; Kimoto, T. Space-Modulated Junction Termination Extension for Ultrahigh-Voltage p-i-n Diodes in 4H-SiC. *IEEE Trans. Electron. Devices* **2012**, *59*, 414–418. [CrossRef]

15. Deng, X.; Li, L.; Wu, J.; Li, C.; Chen, W. A Multiple-Ring-Modulated JTE Technique for 4H-SiC Power Device with Improved JTE-Dose Window. *IEEE Trans. Electron. Devices* **2017**, *64*, 5042–5047. [CrossRef]

16. Perez, R.; Tournier, D.; Perez-Tomas, A.; Godignon, P.; Mestres, N. Planar edge termination design and technology considerations for 1.7-kV 4H-SiC PiN diodes. *IEEE Trans. Electron. Devices* **2005**, *52*, 2309–2316. [CrossRef]

17. Perez, R.; Mestres, N.; Tournier, D. A highly effective edge termination design for SiC planar high power devices. *Mater. Sci. Forum* **2014**, *457–460*, 1253–1256. [CrossRef]

18. Zhang, F.; Li, X. Research on high-voltage 4H-SiC P-i-N diode with planar edge junction termination techniques. *Chin. Phys. B* **2011**, *20*, 366–371. [CrossRef]

19. Huang, R.; Tao, Y.; Wang, L.; Chen, G.; Bai, S. Development of g10kv 4H-SiC SBD junction extension termination. In Proceedings of the International Conference on Advanced Electronic Science and Technology, Shenzhen, China, 19–21 August 2016. [CrossRef]

20. Sheridan, D.C.; Niu, G.; Cressler, J.D. Design of single and multiple zone junction termination extension structures for SiC power devices. *Solid-State Electron.* **2001**, *45*, 1659–1664. [CrossRef]

21. Sung, W.; Brunt, E.V.; Baliga, B.J.; Huang, A.Q. A new edge termination technique for high-voltage devices in 4H-SiC–multiple-floating-zone junction termination extension. *IEEE Electron. Device Lett.* **2011**, *32*, 880–882. [CrossRef]

22. Ghandi, R.; Buono, B.; Domeij, M.; Zetterling, C.M. High-Voltage 4H-SiC PiN Diodes with Etched Junction Termination Extension. *IEEE Electron. Device Lett.* **2009**, *30*, 1170–1172. [CrossRef]

23. Pâques, G.; Dheilly, N.; Planson, D.; Scharnholz, S. Graded Etched Junction Termination for SiC Thyristors. *Mater. Sci. Forum* **2011**, *679–680*, 457–460. [CrossRef]

24. Huang, C.F.; Hsu, H.C.; Chu, K.W.; Lee, L.H.; Tsai, M.J. Counter-Doped JTE, an Edge Termination for HV SiC Devices with Increased Tolerance to the Surface Charge. *IEEE Trans. Electron. Devices* **2015**, *62*, 354–358. [CrossRef]

25. Kaji, N.; Niwa, H.; Suda, J.; Kimoto, T. Ultrahigh-Voltage SiC p-i-n Diodes with Improved Forward Characteristics. *IEEE Trans. Electron. Devices* **2015**, *62*, 374–381. [CrossRef]

26. Deng, X.; Chen, X.; Li, C.; Shen, H.; Zhang, J. Numerical and experimental study of the mesa configuration in high-voltage 4H–SiC PiN rectifiers. *Chin. Phys. B* **2016**, *25*, 087201. [CrossRef]

27. Baliga, B.J. *Fundamentals of Power Semiconductor Devices*; Springer: New York, NY, USA, 2008; Chapter 3; pp. 91–155. [CrossRef]

28. Kim, S.C.; Bahng, W.; Kang, I.H. Fabrication characteristics of 1.2 kV SiC JBS diode. In Proceedings of the International Conference on Microelectronics, Nis, Serbia & Montenegro, 11–14 May 2008; pp. 181–184.

29. Hiyoshi, T.; Hori, T.; Suda, J. Simulation and Experimental Study on the Junction Termination Structure for High-Voltage 4H-SiC PiN Diodes. *IEEE Trans. Electron. Devices* **2008**, *55*, 1841–1846. [CrossRef]

30. Wang, C.; Yuan, H.; Song, Q. Fabrication of 3.1 kV/10 A 4H-SiC Junction Barrier Schottky Diodes. In Proceedings of the 2015 IEEE 11th International Conference on ASIC (ASICON), Chengdu, China, 3–6 November 2015; pp. 1–3.

31. Lee, C.Y.; Yen, C.T.; Chu, K.W.; Chen, Y.S. A novel 4H-SiC Trench MOS Barrier Schottky rectifier fabricated by a two-mask process. In Proceedings of the International Symposium on Power Semiconductor Devices and ICS, Kanazawa, Japan, 26–30 May 2013; pp. 171–174. [CrossRef]

micromachines

MDPI

Article

Ku-Band 50 W GaN HEMT Power Amplifier Using Asymmetric Power Combining of Transistor Cells

Seil Kim, Min-Pyo Lee, Sung-June Hong and Dong-Wook Kim *

Department of Radio Science and Engineering, Chungnam National University, Daejeon 34134, Korea; ksl4896@naver.com (S.K.); dignitymp20@naver.com (M.-P.L.); hsj_1006@naver.com (S.-J.H.)
* Correspondence: dwkim21c@cnu.ac.kr; Tel.: +82-42-821-6887

Received: 12 October 2018; Accepted: 20 November 2018; Published: 24 November 2018

check for
updates

Abstract: In this paper, we present a Ku-band 50 W internally-matched power amplifier that asymmetrically combines the power transistor cells of the GaN high electron mobility transistor (HEMT) (CGHV1J070D) from Wolfspeed. The amplifier is designed using a large-signal transistor cell model in the foundry process, and asymmetric power combining, which consists of a slit pattern, oblique wire bonding and an asymmetric T-junction, is applied to obtain the amplitude/phase balance of the combined signals at the transistor cell combining position. Input and output matching circuits are implemented using a thin film process on a titanate substrate and an alumina substrate with the relative dielectric constants of 40 and 9.8, respectively. The pulsed measurement of a 330 µs pulse period and 6% duty cycle shows the maximum saturated output power of 57 to 66 W, drain efficiency of 40.3 to 46.7%, and power gain of 5.3 to 6.0 dB at power saturation from 16.2 to 16.8 GHz.

Keywords: Ku-band; GaN high electron mobility transistor (HEMT); power amplifier; asymmetric power combining; amplitude balance; phase balance

1. Introduction

Conventional radar systems used traveling wave tubes, magnetrons, or klystrons to obtain high output power, but they had the disadvantages of a high operating voltage, large size, a short lifetime, and low reliability. Recently, a GaN high electron mobility transistor (HEMT), which features a low operating voltage, easy maintenance, a small form factor, and better reliability, has been widely used in solid-state power amplifiers (SSPAs), a key component of modern radar systems [1–3]. In comparison with other GaAs-based or Si-based transistors, the GaN HEMT has superior electron transport, high breakdown voltage, and high thermal conductivity; therefore, GaN HEMT power amplifiers showing larger output power and better efficiency have been actively studied and published [4–7].

In this paper, we present a Ku-band internally-matched power amplifier, which uses a GaN HEMT (CGHV1J070D) bare die from Wolfspeed (Durham, NC, USA) and a thin film process for input and output matching circuits on two different substrates. For the balance of the amplitude and phase of the signals at the power-dividing and power-combining positions, we apply the asymmetric power combining of transistor cells which uses a slit pattern, oblique wire bonding and asymmetric T-junction in the input and output matching circuits.

2. Power Amplifier Design

2.1. Device Description

In this work, the GaN HEMT, which is fabricated on a SiC substrate with high thermal conductivity, has a 0.25-µm gate length and operates up to 18 GHz. It has a size of 800 µm × 4800 µm and a saturated output power capability of 70 W at the reference plane of its drain pad. Considering output matching loss, we chose this device for a Ku-band 50 W power amplifier. A photograph of its chip and its main performance parameters are shown in Figure 1 and Table 1 [8].

Figure 1. Photograph of a GaN high electron mobility transistor (HEMT) (CGHV1J070D) from Wolfspeed.

Table 1. Main performance parameters of CGHV1J070D.

Parameters	Specifications
Operating frequency	10 MHz—18 GHz
Saturated output power	70 W
Power-added efficiency	60% at 10 GHz
Small-signal gain	17 dB at 10 GHz
Operating voltage	40 V
Size	800 μm × 4800 μm

2.2. Device's Optimal Source and Load Impedances

The GaN HEMT consists of 12 transistor cells and is estimated to have a total gate width of 14.4 mm [8,9]. Because a conventional large-signal model from Wolfspeed has only 4 ports, we cannot control or tune output signals from the transistor cells to achieve the balance of their amplitudes and phases using only the given model. In this work, we use the transistor cell model (r7 model) of the process design kit from the company's foundry service and modify it to include its gate and drain pad effects obtained by 3D electromagnetic simulations.

Figure 2 compares S-parameters and load-pull results of the conventional large-signal CGHV1J070D model and our modified r7 model. Contrary to the former, the latter has 24 ports as input and output ports and shows almost the same DC curve, maximum available gain and stability performance. As seen in Figure 2, the S-parameter and load-pull simulation results also show only a slight difference due to the parasitic gate and drain pad capacitance.

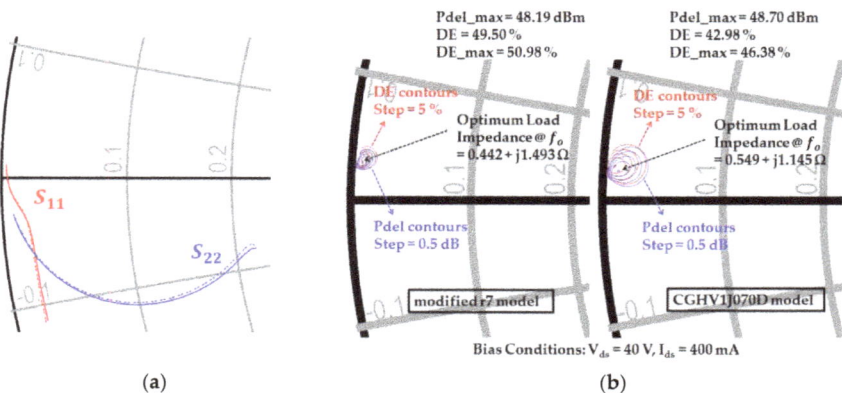

Figure 2. Comparison of S-parameter and load-pull simulation results of the conventional large-signal CGHV1J070D model and our modified r7 model: (**a**) S-parameter simulation; (**b**) load-pull simulation (**left**: modified r7 model, **right**: conventional CGHV1J070D model).

The optimum source impedance and load impedance are obtained by source-pull and load-pull simulation results of our modified r7 model using the Keysight Advanced Design System (ADS) circuit simulator (ADS 2015, Keysight, Santa Rosa, CA, USA). The simulation results show that the optimum source impedance and load impedance are $Z_S = 0.121 - j\,0.251\ \Omega$ and $Z_L = 0.442 + j\,1.493\ \Omega$ at 16.5 GHz, respectively. The obtained optimum impedances are so small that the impedance matching is very challenging in the Ku-band power amplifier design.

2.3. Input and Output Matching Circuit Design

Typically, the optimum source impedance and load impedance of a high-power transistor operating at 10 GHz or above are so small that its impedance matching requires a multi-stage impedance transformer with a high impedance ratio to secure the proper operating bandwidth, which results in a large matching circuit. In the internally-matched power amplifier, which integrates transistor bare dies and thin-film matching circuits in a standard package, a large impedance-transforming matching pattern is not preferable. While maintaining the circuit performance, reducing the size of the matching circuit is essential because of the limited space available in the package. In this work, we use a titanate substrate with a high relative dielectric constant of 40 near the transistor to obtain the impedance matching trace in the low-Q region of the Smith chart and to reduce the size of the matching circuit. The overall input and output matching circuits are fabricated on two different substrates with the relative dielectric constants of 40 and 9.8, respectively. Figure 3 presents a schematic circuit diagram of our Ku-band 50 W GaN HEMT power amplifier.

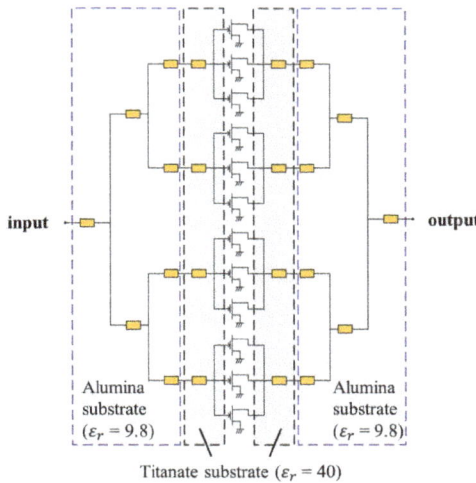

Figure 3. Schematic circuit diagram of the Ku-band 50 W GaN HEMT internally-matched power amplifier.

The output signal of each transistor cell should be combined with the same amplitude and phase to generate high output power efficiently. Although the signal traces in the matching circuit are symmetrically placed at the combining point, the output signals experience different delays due to a combination of various bends, so the power-combining element with the symmetric matching pattern provides each transistor cell with different load impedance, not the optimum load impedance. This results in degraded power performance, rather than an optimized output power. Several design techniques are studied and published to resolve this problem [10–13].

In this work, to obtain amplitude and phase balance at the power-combining point, we apply the asymmetric power combining, which consists of a slit pattern, oblique wire bonding and an asymmetric

T-junction, to the matching circuit patterns on the gate and drain sides. The slit is fabricated on a titanate substrate and the oblique wire bonding and asymmetric T-junction are implemented on an alumina substrate. Figure 4 shows the slit, oblique wire bonding and asymmetric T-junction which are applied to 6 transistor cells. The slit is located on the side of the outer transistor cells (ports 1 to 3), and the T-junction is off-center to the inner transistor cells (ports 4 to 6). In addition to the slit and asymmetric T-junction, the oblique wire bonding improves the maximum amplitude imbalance within 0.3 dB and the maximum phase imbalance within 0.2 degree from 16 GHz to 17 GHz.

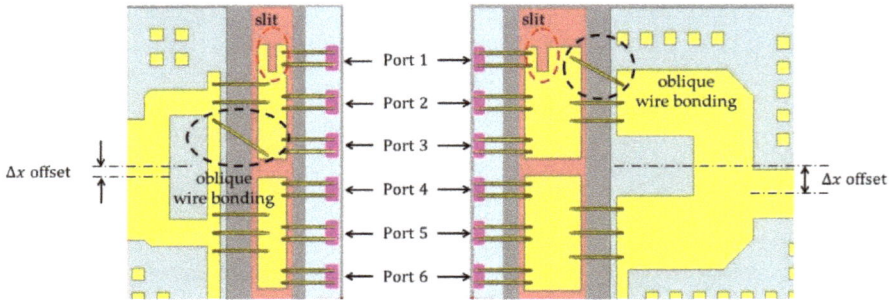

Figure 4. Input and output matching patterns with the slit, oblique wire bonding and asymmetric T-junction.

Figure 5 shows impedance traces on the Smith chart which are seen at several representative positions of the input and output matching circuits. The impedance matching patterns are carefully designed to maintain their traces in the low-Q region of the Smith chart. Figure 6 compares the designed source impedance and load impedance with the optimum source impedance and load impedance. The designed impedance traces make the saturated output power be within the contour plots above 47.6 dBm.

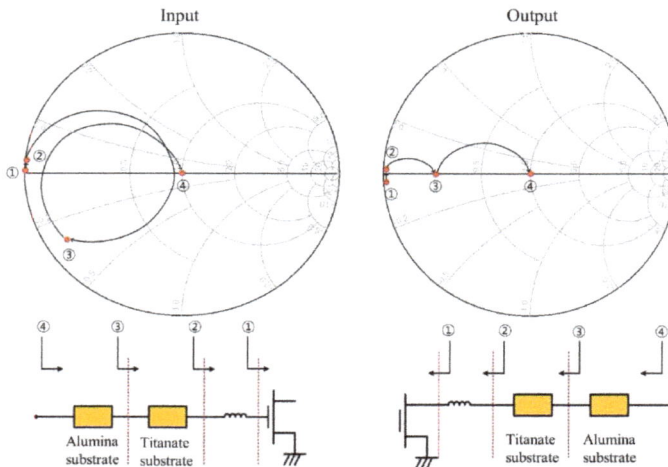

Figure 5. Impedance traces at the representative positions of input and output matching circuits.

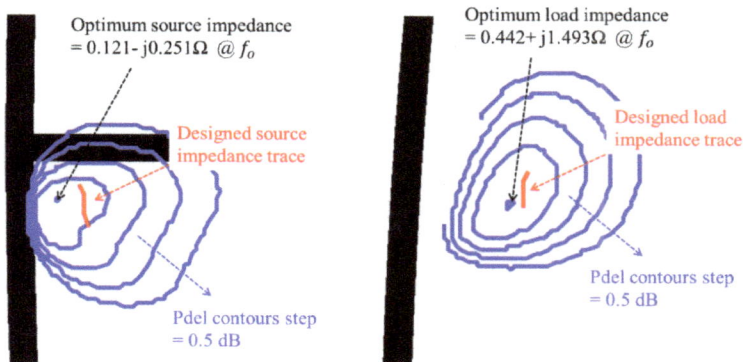

Figure 6. Designed source and load impedance traces overlapped on the simulated source-pull and load-pull contour plots.

3. Fabrication and Measurement

3.1. Power Amplifier Fabrication

A GaN HEMT bare die and 4 input/output matching circuits on the titanate and alumina substrates are attached to a CuW carrier using a eutectic die-attach process and silver epoxy process, to facilitate heat sinking and spreading and interconnect the power amplifier circuit to a printed circuit board (PCB) test circuit using 1-mil wedge bonding. Figure 7 shows the fabricated internally-matched power amplifier.

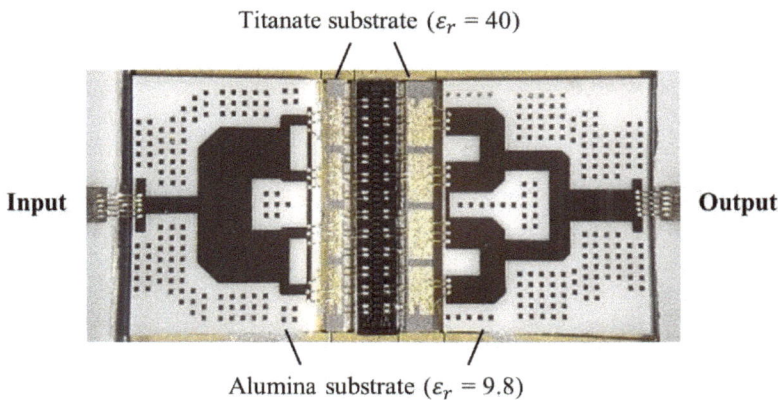

Figure 7. Fabricated internally-matched power amplifier using a GaN bare die and 4 input and output matching substrates.

3.2. Power Amplifier Measurement

The performance of the fabricated GaN HEMT power amplifier was measured under the bias conditions of V_{DS} = 40 V and I_{DS} = 400 mA. Figure 8 compares the simulated (dotted lines) and measured (solid lines) S-parameter results of the power amplifier. The measured small-signal gain (S_{21}) was more than 8.9 dB and the input return loss was better than 7.7 dB in the range of 16.2 to 16.8 GHz, which was in good agreement with the simulated results.

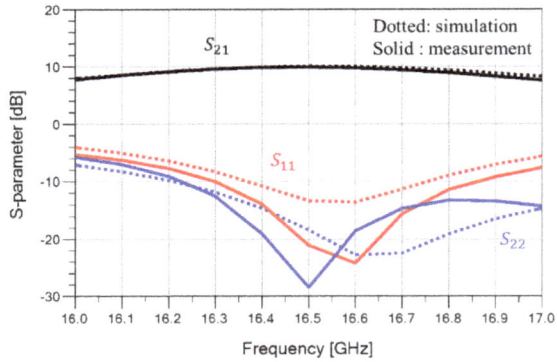

Figure 8. Simulated and measured S-parameter results (simulation: dotted lines, measurement: solid lines).

The output power performance of the fabricated power amplifier was measured from 16.2 to 16.8 GHz and is shown in Figure 9. Figure 9a shows the power gain and drain efficiency with the output power at 16.5 GHz and Figure 9b shows the saturated output power, power gain, and drain efficiency with the input signal frequency of 16.2 to 16.8 GHz, under the pulsed measurement with the pulse period of 330 μs and the duty cycle of 6%. The measured output power performance showed the saturated output power of 48.0 dBm (63.2 W), the power gain of 6 dB, and the drain efficiency of 44.6% at 16.5 GHz. From 16.2 to 16.8 GHz, the measured output power was 47.6 to 48.2 dBm (57 to 66 W), and the drain efficiency was 40.3 to 46.7%, while the power gain was 5.3 to 6.0 dB.

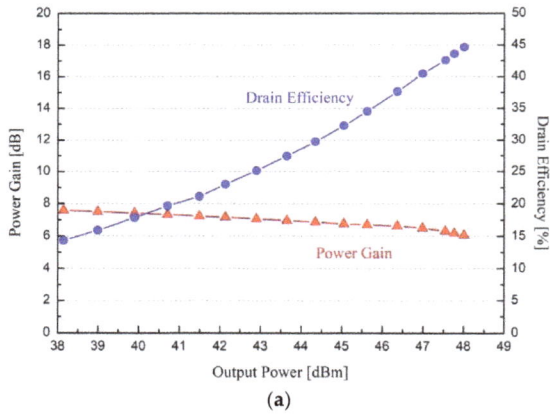

(a)

Figure 9. *Cont.*

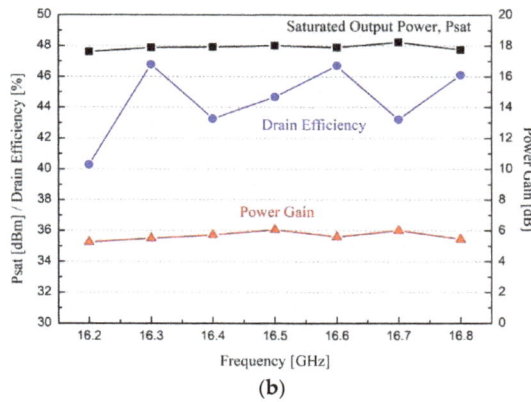

(b)

Figure 9. Measured output power performance results of the fabricated power amplifier at V_{DS} = 40 V and I_{DS} = 400 mA under the pulsed condition: (**a**) power gain and drain efficiency with output power at 16.5 GHz and (**b**) saturated output power, power gain and drain efficiency with input signal frequency of 16.2 to 16.8 GHz.

Table 2 compares the measured output power performance with previously published results of GaN power amplifiers operating in the Ku band. As seen in the table, our internally-matched power amplifier achieves comparable performance or better results in terms of output power and efficiency in comparison with the performance of other GaN power amplifiers.

Table 2. Comparison of our work and previously published Ku-band GaN HEMT power amplifiers.

References	Frequency (GHz)	Power Gain (dB)	P_{sat} (W)	Efficiency @ P_{sat} (%)	Drain Voltage (V)
[14]	16.0	12.8	24.2	22 [1]	31
[15]	16.0~16.5	6.1 [2]	8 [2]	25 [1,2]	8
[16]	16.2~16.8	5.0	50	30	50
[17]	14.0~14.5	5.5	50	21 [1]	40
This work	16.2~16.8	5.3~6.0	57~66	40~47	40

[1] Power-added efficiency; [2] Measured values at P_{1dB} compression point.

4. Conclusions

In this work, we designed and fabricated a Ku-band GaN HEMT internally-matched power amplifier using the asymmetric power-combining of the transistor cells, which utilized a slit pattern, oblique wire bonding and an asymmetric T-junction. The asymmetric power-combining helps to obtain the amplitude and phase balance of the transistor power cells, which can increase the power of combined output signals by combining the power transistor cells in phase. The fabricated power amplifier showed the saturated output power of 57 to 66 W, the power gain of 5.3 to 6.0 dB and the drain efficiency of 40.3 to 46.7% from 16.2 GHz to 16.8 GHz under the pulsed condition. The fabricated power amplifier achieved very competitive performance for the applications of Ku-band radar systems and other high-power transmit/receive systems.

Author Contributions: Conceptualization, S.K. and D.-W.K.; design, S.K.; simulation, S.K. and M.-P.L.; measurement, S.K., S.-J.H. and M.-P.L.; writing, D.-W.K.; supervision and project administration, D.-W.K.

Funding: This work was supported by the Technology Development Program (S2490773) funded by the Ministry of SMEs and Startups (MSS, Korea).

Conflicts of Interest: The authors declare no conflict of interest.

References

1. Brookner, E. Phased array radars-past, present and future. In Proceedings of the 2002 RADAR, Edinburgh, UK, 15–17 October 2002; pp. 104–113.
2. Terry, E. Semiconductor technology trends for phased array antenna power amplifiers. In Proceedings of the 2006 3rd European Radar Conference, Manchester, UK, 13–15 September 2006; pp. 269–272.
3. Kwack, J.Y.; Kim, K.W.; Cho, S. 1kW S-band solid state radar amplifier. In Proceedings of the WAMICON 2011 Conference, Clearwater Beach, FL, USA, 18–19 April 2011; pp. 1–4.
4. Runton, D.W.; Trabert, B.; Shealy, J.B.; Vetury, R. History of GaN: High-power RF gallium nitride (GaN) from infancy to manufacturable process and beyond. *IEEE Microw. Mag.* **2013**, *14*, 82–93. [CrossRef]
5. Pengelly, R.S.; Wood, S.M.; Milligan, J.W.; Sheppard, S.T.; Pribble, W.L. A review of GaN on SiC high electron-mobility power transistors and MMICs. *IEEE Trans. Microw. Theory Tech.* **2012**, *60*, 1764–1783. [CrossRef]
6. Kim, D.W. An output matching technique for a GaN distributed power amplifier MMIC using tapered drain shunt capacitors. *IEEE Microw. Wirel. Compon. Lett.* **2015**, *25*, 603–605.
7. Shin, D.H.; Yom, I.B.; Kim, D.W. 6-GHz-to-18-GHz AlGaN/GaN cascaded nonuniform distributed power amplifier MMIC using load modulation of increased series gate capacitance. *ETRI J.* **2017**, *39*, 737–745. [CrossRef]
8. Wolfspeed, GaN HEMT CGHV1J070D. Available online: www.wolfspeed.com (accessed on 1 June 2017).
9. Maassen, D.; Rautscke, F.; Ohnimus, F.; Schenk, L.; Dalisda, U.; Boeck, G. 70W GaN-HEMT *Ku*-band power amplifier in MIC technology. *IEEE Trans. Microw. Theory Tech.* **2017**, *65*, 1272–1283. [CrossRef]
10. Mori, K.; Nishihara, J.; Utsumi, H.; Inoue, A.; Miyazaki, M. X-band 14W high efficiency internally-matched HFET. In Proceedings of the 2008 IEEE MTT-S International Microwave Symposium, Atlanta, GA, USA, 15–20 June 2008; pp. 315–318.
11. Kimura, M.; Yamauchi, K.; Yamanaka, K.; Noto, H.; Kuwata, E.; Otsuka, H.; Inoue, A.; Kamo, Y.; Miyazaki, M. GaN X-band 43% internally-matched FET with 60 W output power. In Proceedings of the 2008 Asia-Pacific Microwave Conference, Macau, China, 16–20 December 2008; pp. 1–4.
12. Maehara, H.; Uchida, H.; Kosaka, N.; Kuwata, E.; Yamanaka, K.; Nishihara, J.; Kawashima, K.; Nakayama, M. Internally matched GaN FET at C-band with 220 W output power and 56% power added efficiency. In Proceedings of the 2012 Asia Pacific Microwave Conference, Kaohsiung, Taiwan, 4–7 December 2012; pp. 358–360.
13. Nagasaka, M.; Nakazawa, S.; Tanaka, S.; Torii, T.; Imai, S.; Utsumi, H.; Kono, M.; Yamanaka, K.; Fukumoto, H. A Ku-band 100 W power amplifier under CW operation utilizing 0.15 μm GaN HEMT technology. In Proceedings of the 2016 Asia-Pacific Microwave Conference, New Delhi, India, 5–9 December 2016; pp. 1–4.
14. Palmour, J.W.; Sheppard, S.T.; Allen, S.T.; Pribble, W.L.; Smith, T.J.; Ring, Z.; Sumakeris, J.J.; Saxler, A.W.; Milligan, J.W. Wide bandgap semiconductor devices and MMICs for RF power applications. In Proceedings of the International Electron Devices Meeting, Washington, DC, USA, 2–5 December 2001; pp. 385–388.
15. Zhong, S.; Chen, T.; Lin, G.; Li, F. 8-watt internally matched GaAs power amplifier at 16–16.5 GHz. In Proceedings of the 2006 8th International Conference on Solid-State and Integrated Circuit Technology, Shanghai, China, 23–26 October 2006; pp. 887–889.
16. RFHIC, RRPO162168050-05A. Available online: www.rfhic.com (accessed on 23 March 2016).
17. Rautschke, F.; Maassen, D.; Ohnimus, F.; Schenk, L.; Dalisda, U.; Boeck, G. A hybrid 50-W GaN-HEMT Ku-band power amplifier. In Proceedings of the 2016 46th European Microwave Conference, London, UK, 4–6 October 2016; pp. 1079–1082.

micromachines

MDPI

Article

Suppressing the Initial Growth of Sidewall GaN by Modifying Micron-Sized Patterned Sapphire Substrate with H₃PO₄-Based Etchant

Wen-Yang Hsu [1], Yuan-Chi Lian [1], Pei-Yu Wu [1], Wei-Min Yong [1], Jinn-Kong Sheu [2], Kun-Lin Lin [3] and YewChung Sermon Wu [1,*]

[1] Department of Materials Science and Engineering, National Chaio Tung University, Hsinchu 300, Taiwan; j06900397@hotmail.com (W.-Y.H.); kiu741@yahoo.com.tw (Y.-C.L.); a3062123@hotmail.com (P.-Y.W.); cowbell506@yahoo.com.tw (W.-M.Y.)
[2] Department of Photonics, National Cheng Kung University, Tainan City 701, Taiwan; jksheu@mail.ncku.edu.tw
[3] National Nano Device Laboratories, Hsinchu300, Taiwan; kllin@narlabs.org.tw
* Correspondence: sermonwu@faculty.nctu.edu.tw; Tel.: +886-3-513-1555

Received: 3 October 2018; Accepted: 22 November 2018; Published: 26 November 2018

check for
updates

Abstract: Micron-sized patterned sapphire substrates (PSS) are used to improve the performance of GaN-based light-emitting diodes (LEDs). However, the growth of GaN is initiated not only from the bottom c-plane but also from the sidewall of the micron-sized patterns. Therefore, the coalescence of these GaN crystals creates irregular voids. In this study, two kinds of nucleation layers (NL)—ex-situ AlN NL and in-situ GaN NL—were used, and the growth of sidewall GaN was successfully suppressed in both systems by modifying the micron-sized PSS surface.

Keywords: micron-sized patterned sapphire substrate; growth of GaN; sidewall GaN

1. Introduction

High-brightness GaN-based light-emitting diodes (LEDs) are used in a wide variety of applications [1,2]. However, a GaN epitaxial layer usually contains several defects due to the large lattice mismatch and the thermal expansion coefficient difference between GaN and sapphire.

An AlN (or GaN) nucleation layer (NL) is commonly introduced prior to growth of GaN epilayer to overcome this lattice mismatch problem [3,4]. Moreover, micron-sized patterned sapphire substrates (PSS) have been successfully used to reduce these defects and enhance the performance of LEDs [5–12].

When PSS are used, the growth of GaN is initiated not only from the bottom c-plane but also from the sidewall of the micron-sized patterns [13–16]. As the growth time increases, irregular voids are created during the coalescence of these GaN crystals [17].

A GaN NL is usually deposited by metal–organic chemical vapor deposition (MOCVD), and it is called in-situ GaN NL. An AlN NL can be deposited either by MOCVD or sputtered physical vapor deposition (PVD), and they are generally called in-situ AlN NL and ex-situ AlN NL, respectively. It has been found that ex-situ sputtered AlN NL has better GaN quality than in-situ GaN NL and in-situ AlN NL [15,18].

In this study, sulfuric–phosphoric acid was used to modify the micron-sized patterns in order to suppress the growth of sidewall GaN. The effect of this modification on the growth mechanism of GaN was also investigated.

2. Materials and Methods

In this study, commercial dry etching c-plane micron-sized PSS (2.8 μm width and 0.2 μm spacing) was modified. As shown in Figure 1, two kinds of PSS samples were used to investigate the effect of modification of micron-sized PSS patterns on the GaN growth mechanism: (1) RPSS (regular PSS without etching) and (2) PSSE (RPSS etched in sulfuric–phosphoric acid (ratio 3:1) at 270 °C for 30 s). As shown in Figure 1b, 3T {1$\bar{1}$05} facets were observed on the pattern of PSSE [19–23].

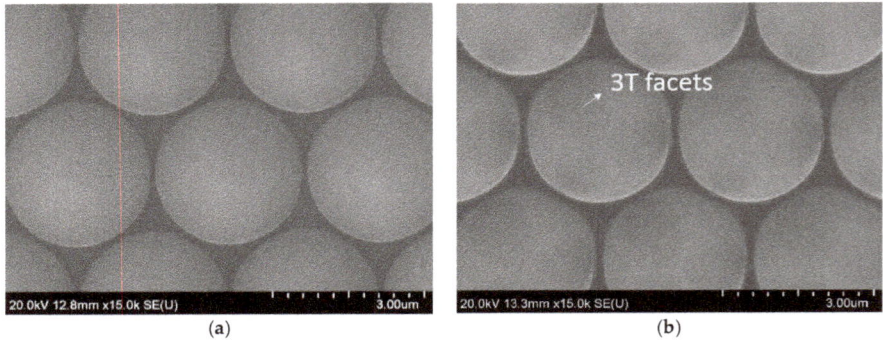

Figure 1. SEM images of (**a**) RPSS (regular patterned sapphire substrates) and (**b**) PSSE (RPSS etched in sulfuric–phosphoric acid).

Two kinds of nucleation layers (NL) were used: (1) ex-situ AlN NL and (2) in-situ GaN NL. To fabricate ex-situ AlN NL, 40 nm AlN was deposited by RF-sputter system using Al target in N_2 gas at 650 °C. As for the in-situ GaN NL, an in-situ 25-nm-thick low-temperature GaN layer was deposited at 550 °C by MOCVD.

As shown in Table 1, four kinds of micron-sized PSS samples were then used to investigate the effect of modification of PSS patterns on the GaN growth mechanism: (1) AlNR (RPSS with AlN NL); (2) AlNE (PSSE with AlN NL); (3) GaNR (RPSS with GaN NL); and (4) GaNE (PSSE with GaN NL).

Table 1. Summary of sample preparation parameters.

Sample	AlNR	AlNE	GaNR	GaNE	AlNOE	GaNOE
Nucleation layers (NL)	AlN	AlN	GaN	GaN	AlN	GaN
PSS substrate	RPSS	PSSE	RPSS	PSSE	PSSO	PSSO

To investigate the GaN epitaxial behavior, high-temperature undoped GaN (HTU-GaN) was grown by MOCVD at 1060 °C with chamber pressure of 200 torr (26,664 Pa) for 2 min.

3. Results

Figure 2 shows the surface morphologies of micron-sized PSS after GaN was grown. The morphologies of bottom GaN (B-GaN) and sidewall GaN (S-GaN) were different. There were two kinds of B-GaN: (1) B3-GaN (GaN grown among three micron-sized patterns) and (2) B2-GaN (GaN grown between two patterns). Two kinds of S-GaN were found: (1) S3-GaN (with AlN as NL; Figure 2a) and (2) S6-GaN (with GaN as NL; Figure 2c). To measure the thicknesses of B-GaN and S-GaN, cross-sectional SEM was carried out by focused ion beam (FIB) cutting along the dash lines as shown in Figure 2a,c. Some of the related images are shown in Figure 3, and the measured maximum thicknesses are summarized in Table 2.

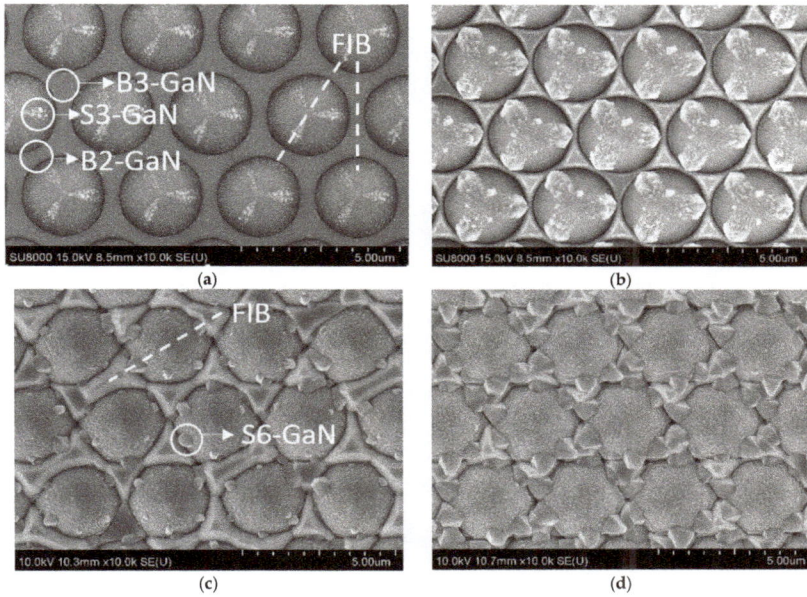

Figure 2. SEM images of GaN grown on (**a**) AlNR, (**b**) AlNE, (**c**) GaNR, and (**d**) GaNE.

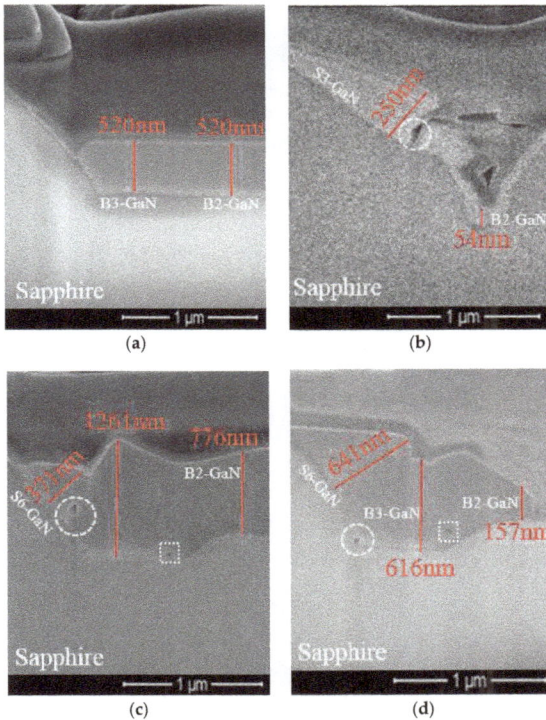

Figure 3. Cross-sectional SEM images from Figure 2. (**a**) AlNR, (**b**) AlNE, (**c**) GaNR, and (**d**) GaNE.

Table 2. The measured maximum thicknesses of GaN.

Thickness	GaN Type	AlNR	AlNE	AlNOE	GaNR	GaNE	GaNOE
	$H_{B3\text{-}GaN}$	520	136	540	1261	616	969
	$H_{B2\text{-}GaN}$	520	54	540	776	157	951
Thickness (nm)	$H_{S3\text{-}GaN}$	74	250	0	0	0	0
	$H_{S6\text{-}GaN}$	0	0	0	371	641	0

4. Discussion

A simple treatment of the MOCVD thin-film growth kinetic involves mass transport and reaction [24,25]. It is reasonable to assume that the mass transport was the same for all the samples as GaN was grown in the same conditions.

In a reaction between A and B to give products C and D, the following applies according to the balance equation:

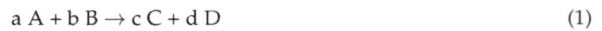

$$a\,A + b\,B \rightarrow c\,C + d\,D \tag{1}$$

The reaction is related to the reactant concentrations in the following way:

$$\text{Rate} = K\,[A]^x[B]^y \tag{2}$$

where K is the rate constant; the numbers x and y are partial orders of reaction.

In this case, there were four surface reaction constants: (1) sidewall with AlN (K_{SAlN}), (2) bottom with AlN (K_{BAlN}), (3) sidewall with GaN (K_{SGaN}), and (4) bottom with GaN (K_{BGaN}).

4.1. Ex-Situ AlN as NL

When ex-situ AlN NL was used, as shown in Figure 2a, two kinds of GaN were found on AlNR. B-GaN was initiated from the bottom c-plane as expected, while S-GaN (S3-GaN) was from sidewall surfaces, which has been reported earlier [14,15,26,27]. Both B-GaN and S3-GaN were Wurtzite structures. The orientation relationship between GaN (including B-GaN and S3-GaN) and sapphire was established as $(0001)_{GaN}$ // (0001) sapphire and $[1\bar{1}00]_{GaN}$ // $[11\bar{2}0]_{sapphire}$.

As shown in Figure 3a, no void was found among GaN crystals as there was no coalescence yet between S-GaN and B-GaN.

Table 2 and Figure 3a show that the maximum thicknesses of B3-GaN ($H_{B3\text{-}GaN}$) and B2-GaN ($H_{B2\text{-}GaN}$) of AlNR were around 520 nm, which was much thicker than that of S3-GaN ($H_{S3\text{-}GaN}$, 74 nm), indicating that K_{BAlN} was much greater than K_{S3AlN} [26].

However, with the modification of PSS patterns (AlNE), $H_{S3\text{-}GaN}$ of AlNE did not diminish but increased. As shown in Figures 2 and 3, and Table 2, compared with AlNR, the $H_{S3\text{-}GaN}$ of AlNE increased from 74 to 250 nm. At the same time, $H_{B3\text{-}GaN}$ decreased from 520 to 136 nm, and $H_{B2\text{-}GaN}$ decreased from 520 to 54 nm.

Moreover, as shown in Figure 3b, irregular voids (circled with dashed lines) were observed between S3-GaN and B2-GaN. These voids were created during the coalescence of GaN crystals [17].

This observation suggested that instead of reducing the reaction constant of S3-GaN (K_{S3AlN}), modification of PSS patterns (AlNE) enhanced K_{SAlN} and reduced K_{BAlN}. As K_{BAlN} should be a constant, we believe this K_{BAlN} reduction should have been caused by the change in the area of the bottom c-plane.

Figure 4 is the high magnifications of (a) RPSS and (b) PSSE. In addition to sidewall facets, an extra six 6B {$3\bar{4}17$} facets were found on the bottom of patterns of PSSE [28], as shown in Figure 4b. The appearance of 6B facets reduced the bottom c-plane fraction as determined by estimating the B3 (B2) vs. total area on the SEM images.

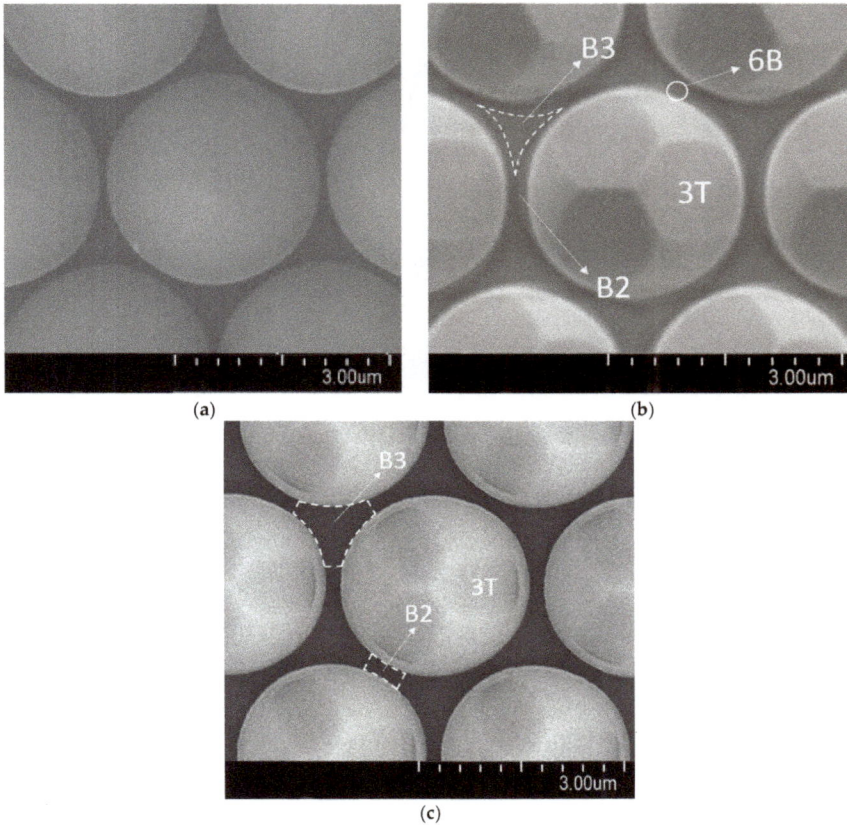

Figure 4. The high magnification SEM images of (**a**) RPSS, (**b**) PSSE, and (**c**) PSSO.

Compared with RPSS, B3 fraction of PSSE reduced from 18% to 12%, while B2 fraction reduced from 3% to 0 %. This reduction of bottom c-plane made epitaxy of GaN film on PSSE very difficult [29]. Consequently, it appeared that the K_{BAlN} of AlNE was much less than that of AlNR.

In addition, compared with AlNR, H_{S3-GaN} of AlNE increased from 74 to 250 nm, as shown in Figures 2 and 3 and Table 2. This is because the consuming of the reactants in front of the bottom c-plane can affect reactant concentrations in front of the sidewall. As shown in Figure 4, the distance between the sidewall and the bottom c-plane was only around 1 μm. As the growth rate of AlNE was much smaller than that of AlNR, the formation of B-GaN of AlNE would consume only a small portion of the reactants. As a result, the reactant concentrations in front of the sidewall of AlNE were increased. Consequently, H_{S3-GaN} of AlNE was thicker than that of AlNR.

4.2. In-Situ GaN as NL

When in-situ GaN NL was used, as shown in Figure 2c, two kinds of GaN were found on AlNR: (1) B-GaN and (2) S6-GaN [13–16]. They were both Wurtzite structures, and the orientation relationship between GaN and sapphire was established as $(0001)_{GaN}$ // (0001) sapphire and $[1\bar{1}00]_{GaN}$ // $[11\bar{2}0]_{sapphire}$.

Table 2 and Figure 3c show that $H_{B3\text{-}GaN}$ and $H_{B2\text{-}GaN}$ of GaNR were 1261 nm and 776 nm, respectively. Their thicknesses were much greater than $H_{S6\text{-}GaN}$ (371 nm) [14,27]. In other words, K_{BGaN} was much greater than K_{S6GaN}.

We also found that modification of PSS patterns (PSSE) did not diminish the growth of sidewall GaN (S6-GaN). Compared with GaNR, $H_{S6\text{-}GaN}$ of GaNE increased from 371 to 614 nm. At the same time, $H_{B3\text{-}GaN}$ decreased from 1261 to 616 nm, and $H_{B2\text{-}GaN}$ decreased from 776 to 157 nm, as shown in Figure 2 and Table 2. We believe that these thickness changes were also due to the reduction in the bottom c-plane of GaNE.

In both cases, beside voids between S3-GaN and B2-GaN (circled with dashed lines), voids were also found between B3-GaN and B2-GaN (squared with dashed lines). These voids were created during the coalescence of GaN crystals [17].

4.3. Bottom C-Plane Protection

To avoid the reduction of the bottom c-plane areas of micron-sized PSS, the bottom c-plane was protected by SiO_2 and then etched with sulfuric–phosphoric acid. This was designated as PSSO. Figure 5 shows the PSSO fabrication processes. Micron-sized RPSS was first deposited with 200-nm-thick SiO_2 film (Figure 5a). A photoresist (PR) layer was spun onto the surface to protect the bottom oxide (B-OX). Sidewall oxide (S-OX) and PR were then removed, as shown in Figure 5b,c. Samples were etched in sulfuric–phosphoric acid at 270 °C for 30 s. B-oxide was then removed (Figure 5e).

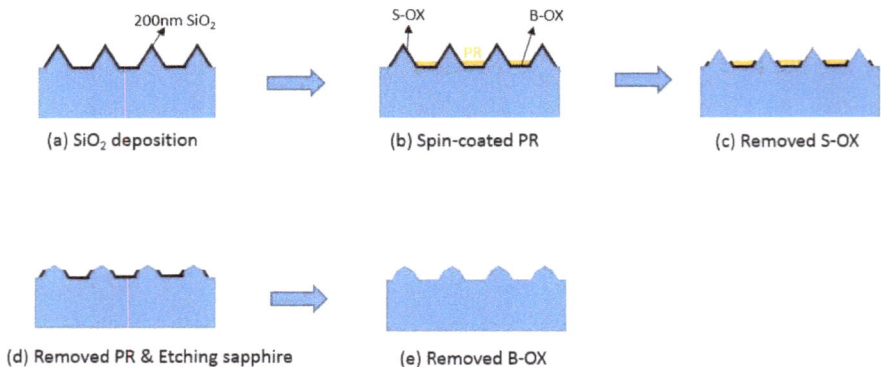

Figure 5. The flow charts of PSSO fabrication processes.

Figure 4c is the high magnification of the PSSO surface. Only sidewall 3T facets were found, and no 6B facets were observed. Compared with RPSS, B3 fraction and B2 fraction were the same as those of RPSS. There was no reduction in the bottom c-plane areas of micron-sized PSSO.

Two kinds of PSSO samples were then fabricated to investigate the growth mechanism of GaN: (1) AlNOE (PSSO with AlN NL) and (2) GaNOE (PSSO with GaN NL).

As shown in Table 2 and Figure 6, no S3-GaN was grown from AlNOE, and no S6-GaN was grown from GaNOE either. In both case, no void was found among GaN crystals, as shown in Figure 7.

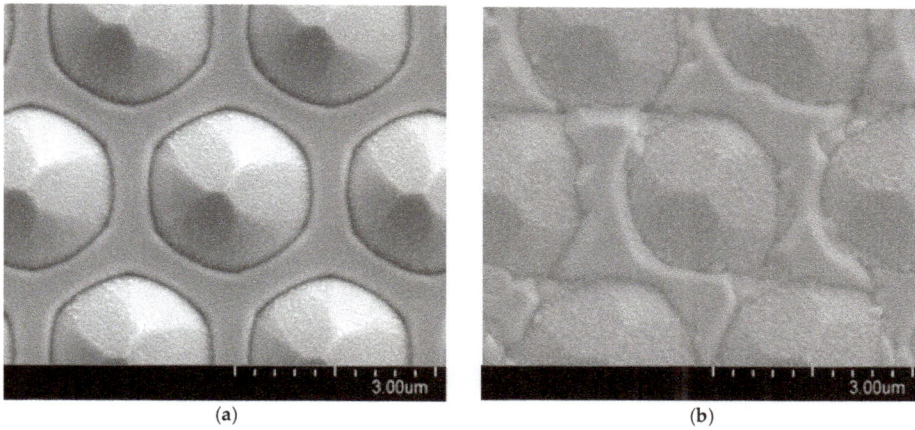

Figure 6. SEM images of GaN grown on (**a**) AlNOE and (**b**) GaNOE.

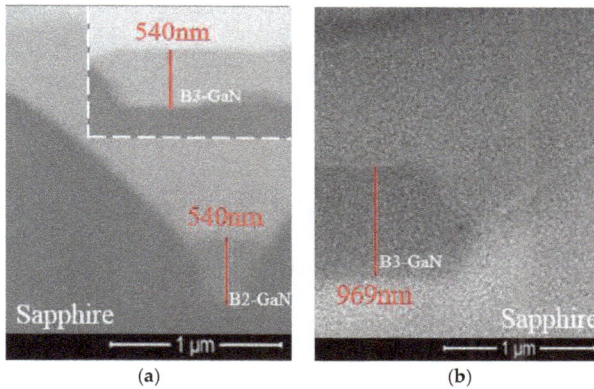

Figure 7. Cross-sectional SEM images from Figure 6. (**a**) AlNOE and (**b**) GaNOE.

5. Conclusions

In this study, the growth of sidewall GaN was successfully suppressed by modifying the surface of micron-sized PSS. Sulfuric–phosphoric acid was used to modify the surface of dry etching c-plane PSS. Two kinds of nucleation layers—ex-situ AlN NL and in-situ GaN NL—were introduced prior to growth of GaN epilayer.

After etching, three 3T $\{1\bar{1}05\}$ facets were found on the pattern sidewall. At the same time, six 6B $\{3\bar{4}17\}$ facets were observed on the bottom of the patterns. The appearance of 6B facets reduced the bottom c-plane fraction, which made epitaxy of GaN on bottom c-plane very difficult. Consequently, instead of reducing the growth of sidewall GaN, this modification appeared to enhance the growth of GaN from the sidewall of the patterns.

A 200-nm-thick SiO_2 film was used to protect the bottom c-plane areas. After etching, only sidewall 3T facets were observed, and no 6B facet appeared. The bottom c-plane areas did not reduce. As a result, sidewall GaN was successfully suppressed in both NL systems.

Author Contributions: Conceptualization, W.-Y.H. and Y.S.W.; Formal analysis, Y.-C.L., P.-Y.W., W.-M.Y., K.-L.L. and Y.S.W.; Investigation, W.-Y.H., Y.-C.L., P.-Y.W., W.-M.Y., J.-K.S. and Y.S.W.; Methodology, W.-Y.H.; Project administration, Y.S.W.; Writing—original draft, W.-Y.H.; Writing—review & editing, Y.S.W.

Funding: This research was funded by (1) the National Science Council of the Republic of China under Grant No. MOST105-2221-E-009-051-MY3, (2) the Ministry of Science and Technology, under Grant No. MOST107-3017-F-009-002, and (3) the "Center for Semiconductor Technology Research" from The Featured Areas Research Center Program within the framework of the Higher Education Sprout Project by the Ministry of Education (MOE) in Taiwan.

Acknowledgments: Technical support from the National Nano Device Laboratory, Center for Nano Science and Technology, Nano Facility Center, and Semiconductor Laser Technology Laboratory of the National Chiao Tung University are also acknowledged.

Conflicts of Interest: The authors declare no conflict of interest.

References

1. Mukai, T.; Nagahama, S.; Sano, M.; Yanamoto, T.; Morita, D.; Mitani, T.; Narukawa, Y.; Yamamoto, S.; Niki, I.; Yamada, M.; et al. Recent progress of nitride-based light emitting devices. *Phys. Status Solidi A* **2003**, *200*, 52–57. [CrossRef]
2. Mukai, T.; Yamada, M.; Nakamura, S. Characteristics of InGaN-based UV/blue/green/amber/red light-emitting diodes. *Jpn. J. Appl. Phys.* **1999**, *38*, 3976–3981. [CrossRef]
3. Amano, H.; Sawaki, N.; Akasaki, I.; Toyoda, Y. Metalorganic vapor phase epitaxial growth of a high quality GaN film using an AlN buffer layer. *Appl. Phys. Lett.* **1986**, *48*, 353–355. [CrossRef]
4. Nakamura, S.; Mukai, T.; Senoh, M. In situ monitoring and Hall measurements of GaN grown with GaN buffer layers. *J. Appl. Phys.* **1992**, *71*, 5543–5549. [CrossRef]
5. Wuu, D.S.; Wang, W.K.; Wen, K.S.; Huang, S.C.; Lin, S.H.; Horng, R.H.; Yu, Y.S.; Pan, M.H. Fabrication of pyramidal patterned sapphire substrates for high-efficiency InGaN-based light emitting diodes. *J. Electrochem. Soc.* **2006**, *153*, G765–G770. [CrossRef]
6. Lee, J.H.; Oh, J.T.; Kim, Y.C.; Lee, J.H. Stress reduction and enhanced extraction efficiency of GaN-based LED grown on cone-shape-patterned sapphire. *IEEE Photonics Technol. Lett.* **2008**, *20*, 1563–1565. [CrossRef]
7. Tsai, C.H.; Ma, M.H.; Yin, Y.F.; Li, H.W.; Lai, W.C.; Huang, J. Nanoscale epitaxial lateral overgrowth of GaN-based light-emitting diodes on an AlN nanorod-array template. *IEEE J. Quantum Electron.* **2015**, *51*, 2006–2009. [CrossRef]
8. Lin, H.C.; Lin, R.S.; Chyi, J.I.; Lee, C.M. Light output enhancement of InGaN light-emitting diodes grown on Masklessly etched sapphire substrates. *IEEE Photonics Technol. Lett.* **2008**, *20*, 1621–1623. [CrossRef]
9. Hsieh, C.-Y.; Lin, B.-W.; Cho, H.-J.; Wang, B.-M.; Sermon Wu, Y. Investigation of GaN Films Grown on Liquid-Phase Deposited SiO2 Nanopatterned Sapphire Substrates. *ECS J. Solid State Sci. Technol.* **2012**, *1*, Q35–Q37. [CrossRef]
10. Lee, Y.J.; Hsu, T.C.; Kuo, H.C.; Wang, S.C.; Yang, Y.L.; Yen, S.N.; Chu, Y.T.; Shen, Y.J.; Hsieh, M.H.; Jou, M.J.; et al. Improvement in light-output efficiency of near-ultraviolet InGaN-GaN LEDs fabricated on stripe patterned sapphire substrate. *Mater. Sci. Eng. B* **2005**, *122*, 184–187. [CrossRef]
11. Feng, Z.H.; Lau, K.M. Enhanced luminescence from GaN-based blue LEDs grown on grooved sapphire substrates. *IEEE Photonics Technol. Lett.* **2005**, *17*, 1812–1814. [CrossRef]
12. Lee, Y.J.; Hsu, T.C.; Kuo, H.C.; Wang, S.C.; Yang, Y.L.; Yen, S.N.; Chu, Y.T.; Shen, Y.J.; Hsieh, M.H.; Jou, M.J.; et al. Enhancing the output power of GaN-based LEDs grown on wet-etched patterned sapphire substrates. *IEEE Photonics Technol. Lett.* **2006**, *18*, 1152–1154. [CrossRef]
13. Zhou, S.; Hu, H.; Liu, X.; Liu, M.; Ding, X.; Gui, C.; Liu, S.; Guo, L.J. Comparative study of GaN-based ultraviolet LEDs grown on different-sized patterned sapphire substrates with sputtered AlN nucleation layer. *Jpn. J. Appl. Phys.* **2017**, *56*, 111001. [CrossRef]
14. He, C.; Zhao, W.; Zhang, K.; He, L.; Wu, H.; Liu, N.; Zhang, S.; Liu, X.; Chen, Z. High-Quality GaN Epilayers Achieved by Facet-Controlled Epitaxial Lateral Overgrowth on Sputtered AlN/PSS Templates. *ACS Appl. Mater. Interfaces* **2017**, *9*, 43386–43392. [CrossRef] [PubMed]
15. Hu, H.; Zhou, S.; Liu, X.; Gao, Y.; Gui, C.; Liu, S. Effects of GaN/AlGaN/Sputtered AlN nucleation layers on performance of GaN-based ultraviolet light-emitting diodes. *Sci. Rep.* **2017**, *7*, 1–10. [CrossRef] [PubMed]
16. Chen, S.-W.; Yang, Y.; Wen, W.-C.; Li, H.; Lu, T.-C. Significant improvement of GaN crystal quality with ex-situ sputtered AlN nucleation layers. *Int. Soc. Opt. Photonics* **2016**, *9768*, 97681D. [CrossRef]

17. Chang, L.C.; Chen, Y.A.; Kuo, C.H. Spatial correlation between efficiency and crystal structure in GaN-based light-emitting diodes prepared on high-aspect ratio patterned sapphire substrate with sputtered AlN nucleation layer. *IEEE Trans. Electron Devices* **2014**, *61*, 2443–2447. [CrossRef]
18. Lai, W.C.; Yen, C.H.; Yang, Y.Y.; Wang, C.K.; Chang, S.J. GaN-Based ultraviolet light emitting diodes with Ex situ sputtered AlN nucleation layer. *IEEE/OSA J. Disp. Technol.* **2013**, *9*, 895–899. [CrossRef]
19. Chen, Y.-C.; Hsiao, F.-C.; Lin, B.-W.; Wang, B.-M.; Sermon Wu, Y.; Hsu, W.-C. The Formation and the Plane Indices of Etched Facets of Wet Etching Patterned Sapphire Substrate. *J. Electrochem. Soc.* **2012**, *159*, D362. [CrossRef]
20. Shen, J.; Zhang, D.; Wang, Y.; Gan, Y. AFM and SEM Study on Crystallographic and Topographical Evolution of Wet-Etched Patterned Sapphire Substrates (PSS). *ECS J. Solid State Sci. Technol.* **2017**, *6*, R24–R34. [CrossRef]
21. Shen, J.; Zhang, D.; Wang, Y.; Gan, Y. AFM and SEM Study on Crystallographic and Topographical Evolution of Wet-Etched Patterned Sapphire Substrates (PSS). *ECS J. Solid State Sci. Technol.* **2017**, *6*, R122–R130. [CrossRef]
22. Shen, J.; Zhang, D.; Wang, Y.; Gan, Y. AFM and SEM Study on Crystallographic and Topographical Evolution of Wet-Etched Patterned Sapphire Substrates (PSS). *ECS J. Solid State Sci. Technol.* **2017**, *6*, R163–R169. [CrossRef]
23. Lee, Y.J.; Chiu, C.H.; Kuo, H.C.; Lu, T.C.; Wang, S.C.; Ng, K.W.; Lau, K.M.; Yang, Z.P.; Chang, S.P.; Lin, S.Y. Simultaneously enhancing internal and extraction efficiencies of GaN-based light emitting diodes via chemical-wet-etching patterned-sapphire-substrate (CWE-PSS). *ECS Trans.* **2007**, *11*, 225–230. [CrossRef]
24. Plummer, J.D.; Deal, M.D.; Griffin, P.B. *Silicon VLSI Technology Fundamentals, Practice and Modeling*, 1st ed.; Prentice Hall: Upper Saddle River, NJ, USA, 2000; p. 514.
25. Ravasio, S.; Momose, T.; Fujii, K.; Shimogaki, Y.; Sugiyama, M.; Cavallotti, C. Analysis of the gas phase kinetics active during GaN deposition from NH_3 and $Ga(CH_3)_3$. *J. Phys. Chem. A* **2015**, *119*, 7858–7871. [CrossRef] [PubMed]
26. Wu, P.-Y.; Li, J.-H.; Hsu, L.-H.; Huang, C.-Y.; Cheng, Y.-J.; Kuo, H.-C.; Wu, Y.S. Effect of Sputtered AlN Location on the Growth Mechanism of GaN. *ECS J. Solid State Sci. Technol.* **2017**, *6*, R131–R134. [CrossRef]
27. Shang, L.; Zhai, G.; Mei, F.; Jia, W.; Yu, C.; Liu, X.; Xu, B. The effect of nucleation layer thickness on the structura evolution and crystal quality of bulk GaN grown by a two-step process on cone-patterned sapphire substrate. *J. Cryst. Growth* **2016**, *442*, 89–94. [CrossRef]
28. Hsueh, H.H.; Ou, S.L.; Cheng, C.Y.; Wuu, D.S.; Horng, R.H. Performance of InGaN light-emitting diodes fabricated on patterned sapphire substrates with modified top-tip cone shapes. *Int. J. Photoenergy* **2014**. [CrossRef]
29. Chen, C.-C.; Hsiao, F.C.; Lin, B.-W.; Hsu, W.-C.; Wu, Y.S. Evolution of Bottom c-Plane on Wet-Etched Patterned Sapphire Substrate. *ECS J. Solid State Sci. Technol.* **2013**, *2*, R169–R171. [CrossRef]

micromachines

MDPI

Article

Effect of Dielectric Distributed Bragg Reflector on Electrical and Optical Properties of GaN-Based Flip-Chip Light-Emitting Diodes

Shengjun Zhou [1,2,3,4,*], Haohao Xu [1,2], Mengling Liu [1,2], Xingtong Liu [1,2], Jie Zhao [1,2], Ning Li [1,2] and Sheng Liu [1,3,*]

1   Key Laboratory of Hydraulic Machinery Transients, Ministry of Education, Wuhan University, Wuhan 430072, China; 2017202080015@whu.edu.cn (H.X.); lml0305@whu.edu.cn (M.L.); 2016202080010@whu.edu.cn (X.L.); 2017202080016@whu.edu.cn (J.Z.); 2017282080133@whu.edu.cn (N.L.)
2   Center for Photonic and Semiconductor, School of Power and Mechanical Engineering, Wuhan University, Wuhan 430072, China
3   Research Center of Electronic Manufacturing and Packaging Integration, Institute of Technological Sciences, Wuhan University, Wuhan 430072, China
4   State Key Laboratory of Applied Optics, Changchun Institute of Optics, Fine Mechanics and Physics, Chinese Academy of Sciences, Changchun 130033, China
*   Correspondence: zhousj@whu.edu.cn (S.Z.); victor_liu63@vip.126.com (S.L.); Tel.: +86-027-5085-3293 (S.Z. & S.L.)

Received: 31 October 2018; Accepted: 6 December 2018; Published: 8 December 2018

check for updates

Abstract: We demonstrated two types of GaN-based flip-chip light-emitting diodes (FCLEDs) with distributed Bragg reflector (DBR) and without DBR to investigate the effect of dielectric TiO_2/SiO_2 DBR on optical and electrical characteristics of FCLEDs. The reflector consisting of two single TiO_2/SiO_2 DBR stacks optimized for different central wavelengths demonstrates a broader reflectance bandwidth and a less dependence of reflectance on the incident angle of light. As a result, the light output power (LOP) of FCLED with DBR shows 25.3% higher than that of FCLED without DBR at 150 mA. However, due to the better heat dissipation of FCLED without DBR, it was found that the light output saturation current shifted from 268 A/cm² for FCLED with DBR to 296 A/cm² for FCLED without DBR. We found that the use of via-hole-based *n*-type contacts can spread injection current uniformly over the entire active emitting region. Our study paves the way for application of DBR and via-hole-based *n*-type contact in high-efficiency FCLEDs.

Keywords: flip-chip light-emitting diodes; distributed Bragg reflector; light output power; external quantum efficiency

1. Introduction

The wide bandgap GaN and related materials have been extensively studied and implemented for optoelectronic devices that emit light in the spectrum between ultraviolet and visible light [1–7]. GaN-based light-emitting diodes (LEDs) have been extensively adopted in a number of applications such as high-resolution micro-displays, automotive lighting, optogenetics, visible light communication (VLC), and solid-state lighting. [8–14]. The progress in the LED development has been attributed to significant improvement in device efficiency [15,16]. To further enhance the performance of LEDs, there is a great need to improve both internal quantum efficiency (IQE) and light extraction efficiency (LEE). The improvement of IQE has played a key role in LED development. Specifically, the IQE of the InGaN LEDs have been improved by using the large overlap quantum well concept or the new active material concept [17–20]. Generally, LEDs should be driven at a high current density to obtain higher light output power (LOP), which also inevitably generates a large portion of heat [21]. Increasing the operating current density of LEDs is also an effective method to decrease the carrier lifetime and increase the modulation bandwidth of VLC [22,23]. However, GaN-based top-emitting LEDs grown on sapphire substrate suffer from inferior heat dissipation performance due to the poor thermal conductivity of sapphire substrate [24,25]. Additionally, the LEE of top-emitting LEDs was limited by the absorption of light by opaque metal electrodes and total internal reflection (TIR) of the generated light at the GaN (n = 2.45)/air (n = 1) interface resulting from their very different refractive indices [26–29]. The vertical structure LEDs fabricated on a substrate with high thermal conductivity (such as Cu) can overcome the thermal issues. However, wafer bonding and laser lift-off techniques, which are critical fabrication processes for vertical structure LEDs, suffer from low-yield and high-cost [30–33]. The flip-chip technology was brought up to overcome these problems. The LEE of flip-chip LEDs (FCLEDs) was relatively higher compared with the top-emitting LED because of lower refraction index contrast between the sapphire (n = 1.77) and air (n = 1) [34,35]. The FCLEDs can also avoid light absorption by the opaque metal electrodes because light is extracted through sapphire substrate [36,37]. Furthermore, FCLEDs are commonly bonded to a high thermal conductivity submount such as silicon, resulting in a superior heat dissipation capability and a higher light output saturation current density. Accordingly, the FCLEDs can effectively improve modulation bandwidths of VLC since the FCLEDs can be operated at a higher injection current density as compared to top-emitting LEDs.

Photons generated from InGaN/GaN multiple quantum wells (MQWs) active region of FCLEDs emit in any direction. As a result, a large portion of photons emitted from the active region will be lost, particularly for those photons emitted downward. Thus, depositing a reflector onto *p*-GaN in order to reflect photons emitted downward can significantly enhance the LEE of the FCLEDs [38]. It has been reported that various reflectors, such as metallic mirrors and dielectric distributed Bragg reflectors (DBRs), have been used to enhance the LEE of FCLEDs [39–42]. In addition, highly reflective DBRs are also important for realizing high performance vertical-cavity surface emitting laser [43,44]. The metallic mirrors have high reflectivity in the visible wavelength range. However, metallic mirrors including Al and Ag suffer from inferior ohmic contact behavior and poor adhesion to the *p*-GaN layer. As an alternative to metallic reflector, the dielectric DBR has many advantages over a metallic reflector, such as low optical loss, high reflectance, and high mechanical robustness [45,46].

In this study, indium-tin oxide (ITO) transparent conductive layer combined with dielectric DBR is used as reflective *p*-type ohmic contact for FCLEDs, which leads to a significant reduction in absorption of light by opaque metal electrodes. We investigated the effect of dielectric TiO_2/SiO_2 DBR on the electrical and optical properties of FCLEDs. The dielectric DBR is composed of 14 alternating nanometer-thick layers of silicon dioxide (SiO_2) and titanium dioxide (TiO_2), which demonstrates high reflectance over the wavelength range from 400 nm to 650 nm at normal incidence. As a result, the light output power of FCLED with DBR was 25.3% higher than that of FCLED without DBR at 150 mA. In addition, via-hole-based *n*-type contacts were used to spread injection current uniformly over the entire active emitting region of FCLEDs.

2. Materials and Methods

GaN epitaxial layers were grown on c-plane (0001) patterned sapphire substrate (PSS) by metal organic chemical vapor deposition (MOCVD). The GaN-based LED structure consists of a 20-nm-thick low-temperature GaN nucleation layer, a 3.0-μm-thick undoped GaN buffer layer, a 2.5-μm-thick Si-doped *n*-GaN layer, a 16-pair $In_{0.02}Ga_{0.98}N$ (2.1 nm)/GaN (2.3 nm) superlattice, a 12-pair $In_{0.16}Ga_{0.84}N$ (3 nm)/GaN (12 nm) multiple quantum wells (MQWs), a 20-nm-thick low-temperature *p*-GaN layer, a 45-nm-thick *p*-$Al_{0.15}Ga_{0.85}N$ electron blocking layer, and a 120-nm-thick Mg-doped *p*-GaN layer. The LED wafer was subsequently annealed at 750 °C at N_2 atmosphere to activate Mg acceptor in the *p*-GaN. The peak wavelength of FCLEDs is 465 nm.

Figure 1 shows a schematic illustration of the fabrication processes for FCLED with DBR. The detailed fabrication processes were shown as follows: a. First, an inductively coupled plasma (ICP) etching based on $BCl_3/Cl_2/Ar$ gas chemistry was used to form *n*-type via holes by etching a portion of the *p*-GaN and the InGaN/GaN MQWs to expose the *n*-GaN layer. b. A 200-nm-thick SiO_2 was then deposited on the *p*-GaN layer by plasma enhanced chemical vapor deposition (PECVD), followed by optical photolithography and buffered oxide etch (BOE) wet etching process to form strip-shaped SiO_2 current blocking layer (CBL). c. A 115-nm-thick ITO was deposited on the *p*-GaN as a *p*-type ohmic contact using electronic beam evaporator, followed by thermal annealing at 550 °C under N_2 ambient. d. Next, Cr/Al/Ti/Pt/Au (20 nm/100 nm/50 nm/50 nm/1 μm) metal was deposited on the ITO and *n*-GaN layers to form the *p*- and *n*-electrodes. e. DBR consisting of 14 alternating pairs of TiO_2/SiO_2 was deposited on the ITO by ion beam sputtering. f. Cr/Al/Ti/Pt/Ti/Pt/Au (20 nm/100 nm/50 nm/50 nm/50 nm/50 nm/1 μm) layers were then evaporated into *p*-type via holes and *n*-type via holes as *p*- and *n*-pads. Finally, the LED wafers were thinned down to be about 150 μm and diced into chips with a dimension of 380 μm × 760 μm. The schematic illustration of the FC-LED with DBR is shown in Figure 2. FCLED without DBR was also fabricated for comparison. For device characterization, current–voltage (I–V) characteristics were measured by using a semiconductor parameter analyzer (Keysight B2901A). The light output power (LOP)–current (L-I) characteristics of LED were determined using a calibrated integrating sphere. The light emission images of LED were obtained using a calibrated charge-coupled device (CCD) camera mounted on a microscope.

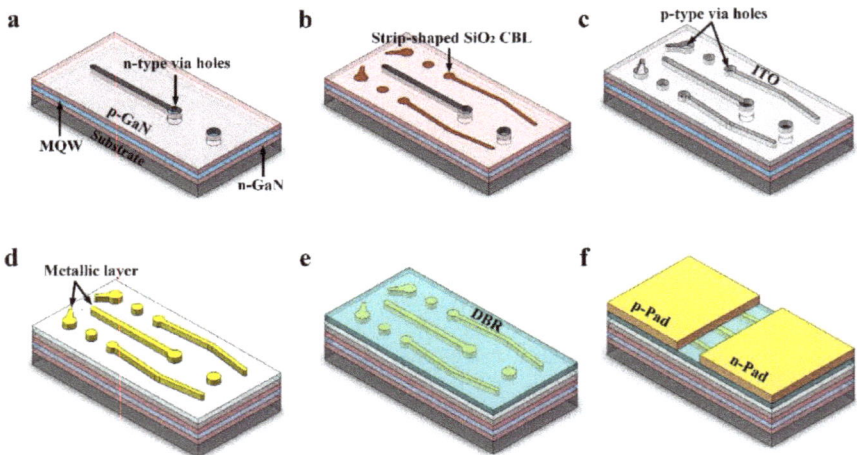

Figure 1. Schematic illustration of the fabrication process for a FCLED with DBR.

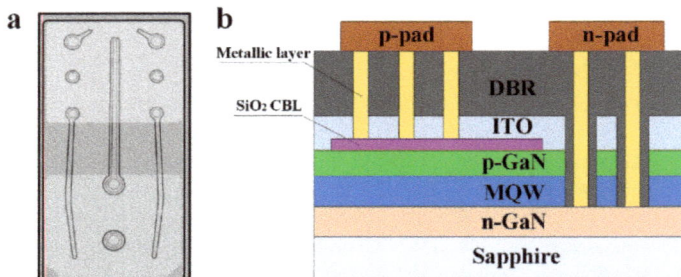

Figure 2. Schematic illustration of a FCLED with DBR: (**a**) Top-view image. (**b**) Cross-section image.

3. Results and Discussion

We used the commercial software, TFCalc, to model the design of a conventional single DBR stack consisting of 14 pairs of TiO_2/SiO_2 dielectric layers optimized for central wavelength at 465 nm. In the simulation, the refractive indices of the SiO_2/TiO_2 were fixed at 1.45/2.55, and the thicknesses of the SiO_2/TiO_2 were fixed at 47.4 nm/79.5 nm. Figure 3a shows reflectance spectra of the 14 pairs of single TiO_2 (47.4 nm) /SiO_2 (79.5 nm) DBR stack as a function of incident angles of light. It was clearly observed that the reflective bandwidth was narrowed and blueshifted toward the short wavelength when the incident angle of light was increased. Figure 3b shows normal-incident reflectance spectra of the single TiO_2/SiO_2 DBR stack optimized for central wavelength of 465 nm, 545 nm, and 620 nm. The thickness of TiO_2/SiO_2 dielectric layers was 47.4 nm/79.5 nm, 64.29 nm/92.73 nm, and 65.0 nm/105.1 nm. As the thickness of the TiO_2/SiO_2 dielectric layers was increased, the reflective bandwidth of the single DBR stack was redshifted. The redshift toward the long wavelength for the single DBR stack with increasing thickness of TiO_2/SiO_2 dielectric layers can counteract the blueshift toward the short wavelength when the incident angle of light increased from the surface normal toward the grazing angle to the DBR stack [47]. To obtain a larger bandwidth of reflectance band and less dependence on incident angles of light, we combined two single DBR stacks into double DBR stacks. Each single TiO_2/SiO_2 DBR stack was optimized for a different central wavelength. The first DBR stack is composed of seven pairs of TiO_2/SiO_2 (47.4 nm/79.5 nm) dielectric layers optimized for a central wavelength at 465 nm; the second DBR stack consists of another seven pairs of TiO_2/SiO_2 (65.0 nm/105.1 nm) dielectric layers optimized for a central wavelength at 620 nm. Figure 3c shows the reflectance spectra of the double DBR stacks as a function of incident angles of light. We find that as the incident light deviates from normal incidence, the blueshift of the double DBR stack is not obvious. This phenomenon indicates that the double DBR stacks exhibit less angular dependence as compared to the single DBR stack. Figure 3d shows the measure reflectance spectra of double DBR stacks. The measured reflectance bandwidth of double DBR stacks was in good agreement with the result of numerical simulation.

Figure 4a shows the top-view SEM images of the FCLED with DBR. The electrode pattern of FCLED without DBR is the same as that of FCLED with DBR. For the FCLEDs with and without DBR, two *p*-contact fingers are finely distributed on both sides of the *n*-contact finger, which can improve the uniformity of current spreading over the active region by reducing lateral current spreading distance between the *p*-type contact and the *n*-type contact. Figure 4b,c show the cross-sectional SEM images of the FCLED with and without DBR, which were obtained by focused ion beam milling along the A-A direction, as shown in Figure 4a. Figure 4d shows magnified cross-sectional SEM image of the FCLED with DBR. In Figure 4d, contact to ITO was obtained by the formation of *p*-type via holes through dielectric TiO_2/SiO_2 DBR. The sheet resistance of 115-nm-thick ITO film (36 Ω/sq) is much larger than that of the as-grown *n*-GaN layer (18 Ω/sq), resulting in severe current crowding at the edge of the *p*-contact electrode. Therefore, a strip-shaped SiO_2 CBL underneath the *p*-electrode was implemented to further improve current spreading of FCLEDs.

Figure 5 shows spatial distribution of light emission intensity images of FCLEDS with and without DBR measured by a calibrated CCD camera. Figure 5a–d show the light emission intensity distribution images of FCLED without DBR at 100 mA, 150 mA, 200 mA and 250 mA. Figure 5e–h show the light emission intensity distribution images of FCLED with DBR at 100 mA, 150 mA, 200 mA and 250 mA. As a higher current density can cause a stronger light emission intensity, the spatial distribution of light emission intensity is closely related to the distribution of current density. We observed that the light emission intensity of FCLEDs increased with the increase of injection current density. The current crowding occurring in both FCLEDs is not obvious at 100 mA, as shown in Figure 5. As the injection current is further increased, the current congregated near the *p*-electrode of FCLEDs, leading to non-uniform light emission intensity in both FCLEDs. It was clearly indicated that the FCLED with DBR exhibited a stronger light emission intensity due to the use of dielectric TiO_2/SiO_2 DBR having high reflectance.

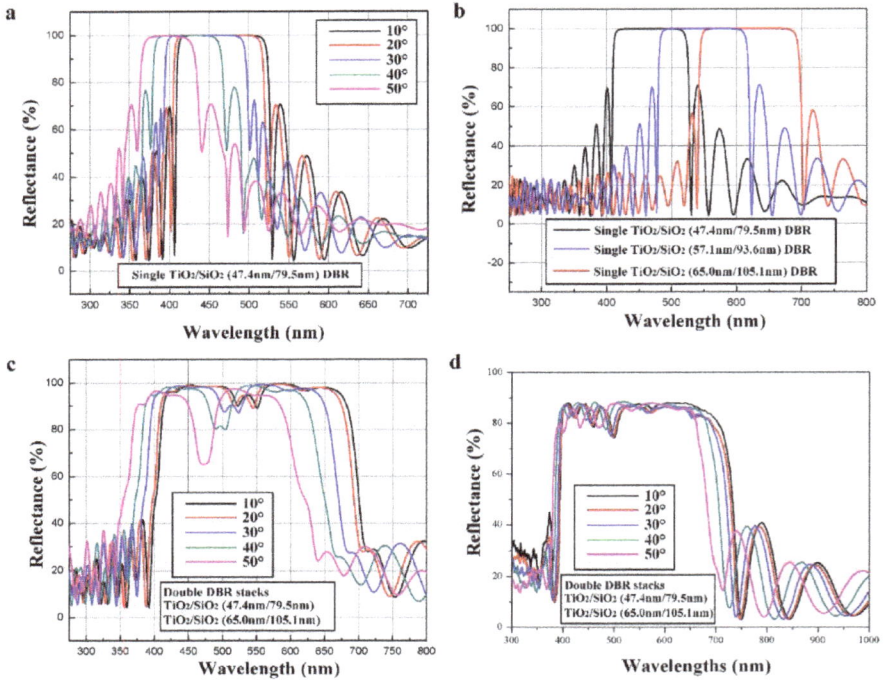

Figure 3. (**a**) Reflectance spectra of the single DBR stack as a function of incident angles of light. (**b**) Normal-incident reflectance spectra of the single TiO_2/SiO_2 DBR stack optimized for a different central wavelength. (**c**) Reflectance spectra of the double DBR stacks as a function of incident angles of light. (**d**) Measured reflectance spectra of double DBR stacks.

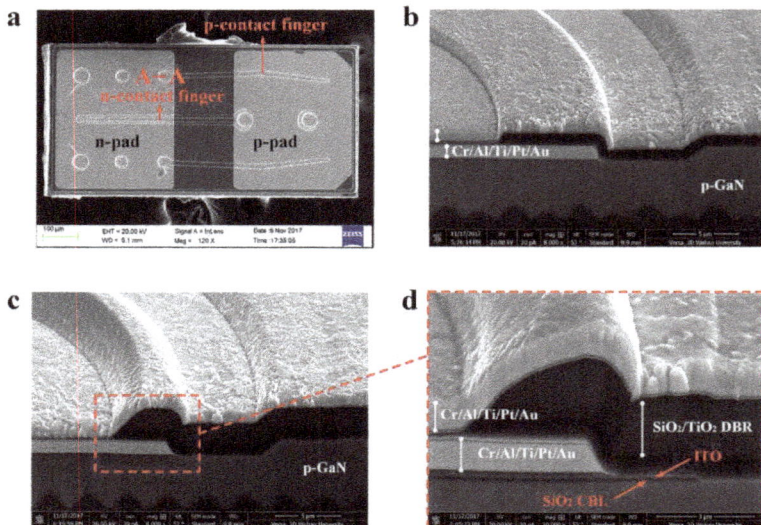

Figure 4. (**a**) Top-view SEM image of the fabricated FCLED with DBR. (**b**) Cross-sectional SEM image of the FCLED without DBR. (**c**) Cross-sectional SEM image of the FCLED with DBR. (**d**) Magnified Cross-sectional SEM image of the FCLED with DBR.

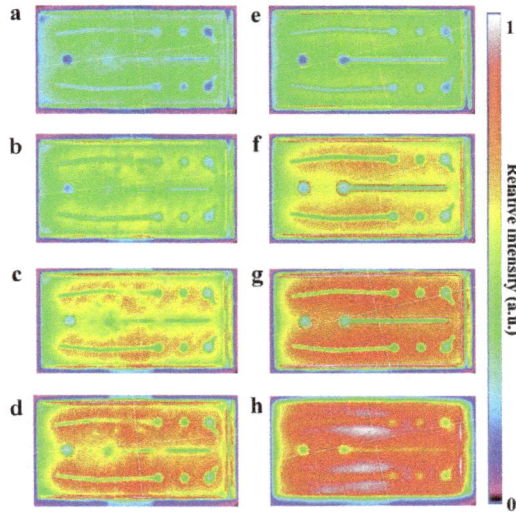

Figure 5. (**a–d**) Light emission intensity distribution images of FCLED without DBR at 100 mA, 150 mA, 200 mA and 250 mA. (**e–h**) Light emission intensity distribution images of FCLED with DBR at 100 mA, 150 mA, 200 mA and 250 mA.

The current versus voltage for the FCLEDs with and without DBR is shown in Figure 6a. At 150 mA, the forward voltages of the FCLEDs with and without DBR were 3.11 V and 3.03 V, respectively. The LOP versus current and external quantum efficiency (EQE) versus current characteristics of the FCLEDs with and without DBR were shown in Figure 6b. At 150 mA, the LOPs of the FCLEDs with DBR and without DBR were 204.6 mW and 152.8 mW, respectively. The LOP of the FCLED with DBR was 25.3% higher than that of the FCLED without DBR at 150 mA. This result can be attributed to the use of DBR having high reflectance in blue light wavelength region. At 150 mA, the corresponding EQEs of the FCLEDs with DBR and without DBR were 49.5% and 39.0%, respectively. The EQE of the FCLED with DBR was 21.2% higher than that of the FCLED without DBR. The light output saturation currents of the FCLED with and without DBR were 875 mA and 1025 mA, respectively. As the dielectric TiO_2/SiO_2 DBR has low thermal conductivity, the FCLED without DBR exhibited higher light output saturation current compared to the FCLED with DBR, owing to a better heat dissipation performance in FCLED without DBR.

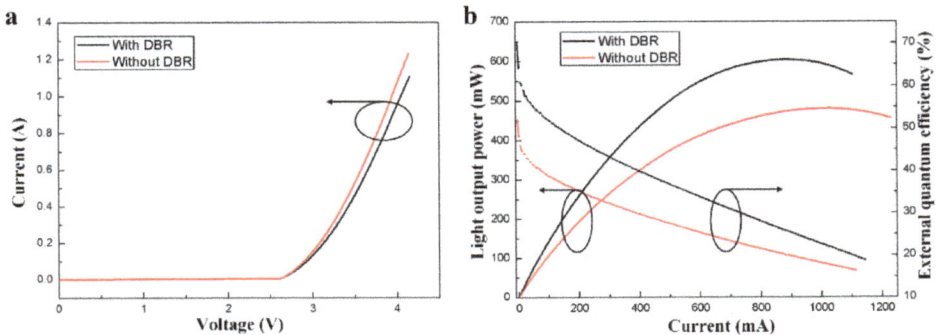

Figure 6. (**a**) Current versus voltage of FCLEDs with and without DBR. (**b**) Light output power versus current and EQE versus current characteristics of FCLEDs with DBR and without DBR.

4. Conclusions

In summary, we have compared the optical and electrical characteristics of FCLEDs with and without DBR. To reduce angular dependence of single DBR stack and increase reflectance bandwidth, we combined two single TiO_2/SiO_2 DBR stack into double DBR stacks. The double DBR stacks exhibited a high reflectance of 97.8% at 465 nm. Additionally, via-hole-based *n*-type contacts were used to improve current spreading of FCLEDs. As a result, the LOP of FCLED with DBR was 25.3% higher than that of the FCLED without DBR at 150 mA. However, owing to low thermal conductivity of dielectric TiO_2/SiO_2 DBR, FCLEDs with DBR exhibited lower light output saturation current as compared to FCLEDs without DBR.

Author Contributions: Conceptualization, S.Z. and S.L.; Data curation, J.Z.; Funding acquisition, S.Z. and S.L.; Investigation, H.X., M.L. and N.L.; Methodology, X.L.; Project administration, S.Z. and S.L.; Supervision, S.Z.; Validation, X.L.; Writing – original draft, S.Z. and H.X.

Funding: This work was funded by the National Key Research and Development Program of China (grant number 2017YFB1104900), the National Natural Science Foundation of China (grant number U1501241, 51675386), and the Hubei Province Science Fund for Distinguished Yong Scholars (grant number 2018CFA091).

Acknowledgments: We acknowledge nanofabrication assistance from the Center for Nanoscience and Nanotechnology at Wuhan University.

Conflicts of Interest: The authors declare no conflict of interest.

References

1. Kang, C.M.; Kong, D.J.; Shim, J.P.; Kim, S.; Choi, S.B.; Lee, J.Y.; Min, J.H.; Seo, D.J.; Choi, S.Y.; Lee, D.S. Fabrication of a vertically-stacked passive-matrix micro-LED array structure for a dual color display. *Opt. Express* **2017**, *25*, 2489–2495. [CrossRef] [PubMed]

2. Tsai, M.S.; Lee, X.H.; Lo, Y.C.; Sun, C.C. Optical design of tunnel lighting with white light-emitting diodes. *Appl. Opt.* **2014**, *53*, H114–H120. [CrossRef] [PubMed]

3. Zhou, S.; Liu, S. Transient measurement of light-emitting diode characteristic parameters for production lines. *Rev. Sci. Instrum.* **2009**, *80*, 095102. [CrossRef] [PubMed]

4. Tan, C.K.; Sun, W.; Wierer, J.J., Jr.; Tansu, N. Effect of interface roughness on Auger recombination in semiconductor quantum wells. *AIP Adv.* **2017**, *7*, 035212. [CrossRef]

5. Hu, H.; Zhou, S.; Liu, X.; Gao, Y.; Gui, C.; Liu, S. Effects of GaN/AlGaN/Sputtered AlN nucleation layers on performance of GaN-based ultraviolet light-emitting diodes. *Sci. Rep.* **2017**, *7*, 44627. [CrossRef] [PubMed]

6. Liu, M.; Zhao, J.; Zhou, S.; Gao, Y.; Hu, J.; Liu, X.; Ding, X. An InGaN/GaN Superlattice to Enhance the Performance of Green LEDs: Exploring the Role of V-Pits. *Nanomaterials* **2018**, *8*, 450. [CrossRef]

7. Hsu, W.Y.; Lian, Y.C.; Wu, P.Y.; Yong, W.M.; Sheu, J.K.; Lin, K.L.; Wu, Y.C.S. Suppressing the initial growth of sidewall GaN by modifying micron-sized patterned sapphire substrate with H_3PO_4-based etchant. *Micromachines* **2018**, *9*, 622. [CrossRef] [PubMed]

8. Dadabayev, R.; Shabairou, N.; Zalevsky, Z.; Malka, D. A visible light RGB wavelength demultiplexer based on silicon-nitride multicore PCF. *Opt. Laser. Technol.* **2019**, *111*, 411–416. [CrossRef]

9. Nikolaevsky, L.; Shchori, T.; Malka, D. Modeling a 1 × 8 MMI Green Light Power Splitter Based on Gallium-Nitride Slot Waveguide Structure. *IEEE Photonic. Tech Lett.* **2018**, *15*, 8. [CrossRef]

10. Chang, S.J.; Kuo, C.H.; Su, Y.K.; Wu, L.W.; Sheu, J.K.; Wen, T.C.; Lai, W.C.; Chen, J.R.; Tsai, J.M. 400-nm InGaN-GaN and InGaN-AlGaN multi quantum well light-emitting diodes. *IEEE J. Sel. Top. Quant.* **2002**, *8*, 744–748. [CrossRef]

11. Ge, P.; Li, Y.; Chen, Z.; Wang, H. LED high-beam headlamp based on free-form microlenses. *Appl. Opt.* **2014**, *53*, 5570–5575. [CrossRef] [PubMed]

12. Zhou, S.; Liu, X.; Yan, H.; Gao, Y.; Xu, H.; Zhao, J.; Quan, Z.; Gui, C.; Liu, S. The effect of nanometre-scale V-pits on electronic and optical properties and efficiency droop of GaN-based green light-emitting diodes. *Sci. Rep.* **2018**, *8*, 11053. [CrossRef] [PubMed]

13. Zhmakin, A.I. Enhancement of light extraction from light emitting diodes. *Phys. Rep.* **2011**, *498*, 189–241. [CrossRef]

14. Krames, M.R.; Shchekin, O.B.; Mueller-Mach, R.; Mueller, G.O.; Zhou, L.; Harbers, G.; Craford, M.G. Status and future of high-power light-emitting diodes for solid-state lighting. *J. Disp. Technol.* **2007**, *3*, 160–175. [CrossRef]

15. DenBaars, S.P.; Feezell, D.; Kelchner, K.; Pimputkar, S.; Pan, C.C.; Yen, C.C.; Tanaka, S.; Zhao, Y.; Pfaff, N.; Farrell, R.; et al. Development of gallium-nitride-based light-emitting diodes (LEDs) and laser diodes for energy-efficient lighting and displays. *Acta Mater.* **2013**, *61*, 945–951. [CrossRef]

16. Tansu, N.; Zhao, H.; Liu, G.; Li, X.H.; Zhang, J.; Tong, H.; Ee, Y.K. III-nitride photonics. *IEEE Photonics J.* **2010**, *2*, 241–248. [CrossRef]

17. Sun, W.; Al Muyeed, S.A.; Song, R.; Wierer, J.J., Jr.; Tansu, N. Integrating AlInN interlayers into InGaN/GaN multiple quantum wells for enhanced green emission. *Appl. Phys. Lett.* **2018**, *112*, 201106. [CrossRef]

18. Arif, R.A.; Ee, Y.K.; Tansu, N. Polarization engineering via staggered InGaN quantum wells for radiative efficiency enhancement of light emitting diodes. *Appl. Phys. Lett.* **2007**, *91*, 091110. [CrossRef]

19. Mitchell, B.; Dierolf, V.; Gregorkiewicz, T.; Fujiwara, Y. Perspective: Toward efficient GaN-based red light emitting diodes using europium doping. *J. Appl. Phys.* **2018**, *123*, 160901. [CrossRef]

20. Tan, C.K.; Borovac, D.; Sun, W.; Tansu, N. InGaN/Dilute-As GaNAs interface quantum well for red emitters. *Sci. Rep.* **2016**, *6*, 19271. [CrossRef]

21. Zhou, S.; Yuan, S.; Liu, S.; Ding, H. Improved light output power of LEDs with embedded air voids structure and SiO$_2$ current blocking layer. *Appl. Surf. Sci.* **2014**, *305*, 252–258. [CrossRef]

22. Teng, D.; Wu, M.; Liu, L.; Wang, G. Size-and current-density-controlled tunable wavelength in GaN-based LEDs for potential dense wavelength-division multiplexing application. *IEEE Wirel. Commun.* **2015**, *22*, 74–79. [CrossRef]

23. Shoresh, T.; Katanov, N.; Malka, D. 1 × 4 MMI visible light wavelength demultiplexer based on a GaN slot-waveguide structure. *Photonic. Nanostruct.* **2018**, *30*, 45–49. [CrossRef]

24. Tawfik, W.Z.; Hyun, G.Y.; Lee, S.J.; Ryu, S.W.; Ha, J.S.; Lee, J.K. Enhanced performance of GaN-based LEDs via electroplating of a patterned copper layer on the backside. *J. Mater. Sci.* **2018**, *53*, 8878–8886. [CrossRef]

25. Han, N.; Cuong, T.V.; Han, M.; Ryu, B.D.; Chandramohan, S.; Park, J.B.; Kang, J.H.; Park, Y.J.; Ko, K.B.; Kim, H.Y.; et al. Improved heat dissipation in gallium nitride light-emitting diodes with embedded graphene oxide pattern. *Nat. Commun.* **2013**, *4*, 1452. [CrossRef] [PubMed]

26. Huh, C.; Lee, K.S.; Kang, E.J.; Park, S.J. Improved light-output and electrical performance of InGaN-based light-emitting diode by microroughening of the p-GaN surface. *J. Appl. Phys.* **2003**, *93*, 9383–9385. [CrossRef]

27. Wierer, J.J.; Krames, M.R.; Epler, J.E.; Gardner, N.F.; Craford, M.G.; Wendt, J.R.; Simmons, J.A.; Sigalas, M.M. InGaN/GaN quantum-well heterostructure light-emitting diodes employing photonic crystal structures. *Appl. Phys. Lett.* **2004**, *84*, 3885–3887. [CrossRef]

28. McGroddy, K.; David, A.; Matioli, E.; Iza, M.; Nakamura, S.; DenBaars, S.; Speck, J.S.; Hu, E.L. Directional emission control and increased light extraction in GaN photonic crystal light emitting diodes. *Appl. Phys. Lett.* **2008**, *93*, 103502. [CrossRef]

29. Zhou, S.; Yuan, S.; Liu, Y.; Guo, L.J.; Liu, S.; Ding, H. Highly efficient and reliable high power LEDs with patterned sapphire substrate and strip-shaped distributed current blocking layer. *Appl. Surf. Sci.* **2015**, *355*, 1013–1019. [CrossRef]

30. Wong, W.S.; Sands, T.; Cheung, N.W.; Kneissl, M.; Bour, D.P.; Mei, P.; Romano, L.T.; Johnson, N.M. In$_x$Ga$_{1-x}$N light emitting diodes on Si substrates fabricated by Pd-In metal bonding and laser lift-off. *Appl. Phys. Lett.* **2000**, *77*, 2822–2824. [CrossRef]

31. Hui, K.N.; Wang, X.H.; Li, Z.L.; Lai, P.T.; Choi, H.W. Design of vertically-stacked polychromatic light-emitting diodes. *Opt. Express* **2009**, *17*, 9873–9878. [CrossRef] [PubMed]

32. Zhou, S.; Liu, S. Study on sapphire removal for thin-film LEDs fabrication using CMP and dry etching. *Appl. Surf. Sci.* **2009**, *255*, 9469–9473. [CrossRef]

33. Lee, S.Y.; Choi, K.K.; Jeong, H.H.; Choi, H.S.; Oh, T.H.; Song, J.O.; Seong, T.Y. Wafer-level fabrication of GaN-based vertical light-emitting diodes using a multi-functional bonding material system. *Semicond. Sci. Tech.* **2009**, *24*, 092001. [CrossRef]

34. Wierer, J.J.; Steigerwald, D.A.; Krames, M.R.; O'shea, J.J.; Ludowise, M.J.; Christenson, G.; Shen, Y.C.; Lowery, C.; Martin, P.S.; Subramanya, S.; et al. High-power AlGaInN flip-chip light-emitting diodes. *Appl. Phys. Lett.* **2001**, *78*, 3379–3381. [CrossRef]

35. Liu, M.; Zhou, S.; Liu, X.; Gao, Y.; Ding, X. Comparative experimental and simulation studies of high-power AlGaN-based 353 nm ultraviolet flip-chip and top-emitting LEDs. *Jpn. J. Appl. Phys.* **2018**, *57*, 031001. [CrossRef]

36. Horng, R.H.; Chuang, S.H.; Tien, C.H.; Lin, S.C.; Wuu, D.S. High performance GaN-based flip-chip LEDs with different electrode patterns. *Opt. Express* **2014**, *22*, A941–A946. [CrossRef] [PubMed]

37. Lv, J.; Zheng, C.; Chen, Q.; Zhou, S.; Liu, S. High power InGaN/GaN flip-chip LEDs with via-hole-based two-level metallization electrodes. *Phys. Status Solidi A* **2016**, *213*, 3150–3156. [CrossRef]

38. Lee, J.R.; Na, S.I.; Jeong, J.H.; Lee, S.N.; Jang, J.S.; Lee, S.H.; Jung, J.J.; Song, T.Y.; Seong, J.O.; Park, S.J. Low resistance and high reflectance Pt/Rh contacts to p-type GaN for GaN-based flip chip light-emitting diodes. *J. Electrochem. Soc.* **2005**, *152*, G92–G94. [CrossRef]

39. Zhou, S.; Liu, X.; Gao, Y.; Liu, Y.; Liu, M.; Liu, Z.; Gui, C.; Liu, S. Numerical simulation and experimental investigation of GaN-based flip-chip light-emitting diodes with highly reflective Ag/TiW and ITO/DBR Ohmic contacts. *Opt. Express* **2017**, *25*, 26615–26627. [CrossRef]

40. Liu, X.; Li, N.; Hu, J.; Gao, Y.; Wang, R.; Zhou, S. Comparative Study of Highly Reflective ITO/DBR and Ni/Ag ohmic Contacts for GaN-Based Flip-Chip Light-Emitting Diodes. *ECS J. Solid State Sci. Technol.* **2018**, *7*, Q116–Q122. [CrossRef]

41. Yamae, K.; Fukshima, H.; Fujimoto, K. Omnidirectional Reflector with Total Internal Reflective Interface for Light Extraction Enhancement of Solid-State Light Source. *Phys. Status Solidi A* **2018**. [CrossRef]

42. Hsu, Y.P.; Chang, S.J.; Su, Y.K.; Chang, C.S.; Shei, S.C.; Lin, Y.C.; Chen, S.C. InGaN/GaN light-emitting diodes with a reflector at the backside of sapphire substrates. *J. Electron. Mater.* **2003**, *32*, 403–406. [CrossRef]

43. Lu, T.C.; Wu, T.T.; Chen, S.W.; Tu, P.M.; Li, Z.Y.; Chen, C.K.; Chen, C.H.; Kuo, H.C.; Wang, S.C.; Zan, H.W.; et al. Characteristics of current-injected GaN-based vertical-cavity surface-emitting lasers. *IEEE J. Sel. Top. Quant.* **2011**, *17*, 1594–1602.

44. Leonard, J.T.; Young, E.C.; Yonkee, B.P.; Cohen, D.A.; Margalith, T.; DenBaars, S.P.; Speck, J.S.; Nakamura, S. Demonstration of a III-nitride vertical-cavity surface-emitting laser with a III-nitride tunnel junction intracavity contact. *Appl. Phys. Lett.* **2015**, *107*, 091105. [CrossRef]

45. Zhou, S.; Zheng, C.; Lv, J.; Gao, Y.; Wang, R.; Liu, S. GaN-based flip-chip LEDs with highly reflective ITO/DBR p-type and via hole-based n-type contacts for enhanced current spreading and light extraction. *Opt. Laser Technol.* **2017**, *92*, 95–100. [CrossRef]

46. Liu, X.; Zhou, S.; Gao, Y.; Hu, H.; Liu, Y.; Gui, C.; Liu, S. Numerical simulation and experimental investigation of GaN-based flip-chip LEDs and top-emitting LEDs. *Appl. Opt.* **2017**, *56*, 9502–9509. [CrossRef]

47. Zhou, S.; Cao, B.; Yuan, S.; Liu, S. Enhanced luminous efficiency of phosphor-converted LEDs by using back reflector to increase reflectivity for yellow light. *Appl. Opt.* **2014**, *53*, 8104–8110. [CrossRef]

micromachines

MDPI

Article

Model Development for Threshold Voltage Stability Dependent on High Temperature Operations in Wide-Bandgap GaN-Based HEMT Power Devices

Huolin Huang [1,2,*], Feiyu Li [1], Zhonghao Sun [1] and Yaqing Cao [1]

1   School of Optoelectronic Engineering and Instrumentation Science, Dalian University of Technology,
     Dalian 116024, China; lifeiyu@mail.dlut.edu.cn (F.L.); sunzhonghao@mail.dlut.edu.cn (Z.S.);
     cyqtmxk@mail.dlut.edu.cn (Y.C.)
2   Key Laboratory for Micro/Nano Technology and System of Liaoning Province, Dalian University of
     Technology, Dalian 116024, China
*   Correspondence: hlhuang@dlut.edu.cn; Tel.: +86-041-184-707865

Received: 23 November 2018; Accepted: 10 December 2018; Published: 14 December 2018

check for
updates

Abstract: Temperature-dependent threshold voltage (V_{th}) stability is a significant issue in the practical application of semi-conductor power devices, especially when they are undergoing a repeated high-temperature operation condition. The V_{th} analytical model and its stability are dependent on high-temperature operations in wide-bandgap gallium nitride (GaN)-based high electron mobility transistor (HEMT) devices that were investigated in this work. The temperature effects on the physical parameters—such as barrier height, conduction band, and polarization charge—were analysed to understand the mechanism of V_{th} stability. The V_{th} analytical model under high-temperature operation was then proposed and developed to study the measurement temperatures and repeated rounds dependent on V_{th} stability. The validity of the model was verified by comparing the theoretical calculation data with the experimental measurement and technology computer-aided design (TCAD) simulation results. This work provides an effective theoretical reference on the V_{th} stability of power devices in practical, high-temperature applications.

Keywords: threshold voltage (V_{th}) stability; gallium nitride (GaN); high electron mobility transistors (HEMTs); analytical model; high-temperature operation

---

1. Introduction

Gallium nitride (GaN)-based high electron mobility transistors (HEMTs) have demonstrated a great potential in the fields of power electronics, mainly owing to their large semi-conductor bandgap (~3.4 eV), low intrinsic carrier concentration, and high-density two dimensional electron gas (2DEG) (>10^{13} cm^{-2}), along with their high electron mobility (>2000 cm^2·V·s^{-1}) at the AlGaN/GaN heterojunction interface [1–3]. In comparison with Si or GaAs-based field-effect transistors (FETs), the wide-bandgap GaN-based devices have lower specific on-resistance and a faster ON/OFF switching speed. Therefore, they are well suited for high power switching applications in renewable energy systems, smart power grids, industrial motors, and the like. In particular, in the last decade, GaN-based HEMTs have attracted significant interest for high-frequency applications because the devices exhibit good noise properties comparable to GaAs-based HEMTs, but with the advantage of having a much higher input power robustness [4–7]. GaN-based HEMTs with a selected short gate length have overcome laterally diffused metal oxide semiconductor (LDMOS)-based transistors for the applications above the L-band owing to their higher frequency capabilities. Furthermore, GaN-based devices are also very promising when employed in various high-temperature environments, such as aerospace turbines and automotive internal combustion engines [8–10]. As a result of the considerable and worldwide attention given to the GaN techniques, they have achieved rapid and remarkable progress.

However, there are still several issues that need to be worked out for the practical application of GaN-based products [11,12]. An urgent task is to investigate the temperature-dependent performance stability, for instance, the device threshold voltage (V_{th}) stability [13]. To date, there has been much research on the temperature-dependent analytical model, focusing on investigating a device's temperature-dependent output current–voltage characteristics and equivalent circuits. The temperature-dependent semi-conductor interface state and electron mobility have been extensively investigated [14–17]. The equivalent circuit modelling of GaN-based HEMTs as a function of ambient temperature has been accurately established, which is meaningful since the equivalent circuit is a very useful tool for circuit designers [18,19]. Besides, some work has studied the V_{th} variation according to the measurement temperature [20–22]. However, there is still a lack of sufficient research on the V_{th} analytical model and, in particular, on its stability when subjected to repeated high temperature operations, which might result in an unrecoverable impact on the material properties, for example the interface trap density.

This paper systematically investigates the V_{th} analytical model and the stability mechanism when subjected to the different measurement temperatures and repeated rounds in GaN-based HEMTs. Considering that a relatively large gate length of 2 μm was employed in the fabricated devices, the direct current (DC) characteristics are analysed in this work. The physical model was developed step-by-step by first building and analysing the V_{th} model in the conventional Schottky gate, with an emphasis on the influence of the physical parameters of the basic device on the V_{th}. The model was then analysed in the metal–insulator–semiconductor (MIS) gate structure, with an emphasis on the influence of the charged interface traps. A series of V_{th} values for the fabricated HEMTs were derived after repeating high-temperature measurements from room temperature to high-temperature for several rounds. The effects of the high temperature on the physical parameters, such as barrier height, conduction band, Fermi level, polarization charge, and interface traps were analysed to understand the V_{th} stability mechanism. Both the experimental measurement of the device and technology computer-aided design (TCAD) simulation work were carried out and the validity of the model was finally verified.

2. Physical Mechanism and Threshold Voltage (V_{th}) Analytical Model

Schottky contact is a basic element in power electronic devices. Therefore, physical modelling work starts with an investigation of the characteristics of simple Schottky-gate HEMT devices. The detailed study of effects of high-temperature on the Schottky-gate devices was carried out to understand the role of the basic physical parameters, such as barrier height, conduction band, Fermi level and polarization charges in the operation at high-temperature. Then, more complicated structures with the MIS gate were involved to analyse the effects of the interface traps beneath the gate, considering that the MIS gate structure has a much higher density of interface traps between the dielectric and AlGaN barrier layer than the Schottky gate. These charged traps could induce an instability of the V_{th} value, especially when the devices were operated under the high-temperature condition [23–25]. The physical mechanisms of the trap density variation with the different operation temperatures and sequences were discussed and the analytical model was proposed and developed.

2.1. Schottky-Gate AlGaN/GaN High Electron Mobility Transistors (HEMTs)

By taking into account the combined effects of the device's physical parameters at different temperatures, such as Schottky barrier, polarization charges, energy band and doping-induced charges, the equation to generally describe the V_{th} of the Schottky-gate devices can be modified and expressed as below [26].

$$V_{th1}(T) = \frac{\varphi_b(T)}{q} - \frac{\Delta E_C(T)}{q} + \frac{E_{F0}(T)}{q} \\ - \frac{qd\sigma_{pol}(T)}{\varepsilon_{AlGaN}} \\ - \frac{qN_d(T)d^2}{2\varepsilon_{AlGaN}} \tag{1}$$

In the equation, φ_b is the contact barrier height between the gate metal and AlGaN barrier layer. ΔE_C is the conduction band offset at the AlGaN/GaN interface. E_{F0} is the energy difference between the intrinsic Fermi level and the conduction band edge of GaN bulk. σ_{pol} is the net polarization charge at the AlGaN/GaN interface. N_d is the doping concentration in the AlGaN layer, and ε_{AlGaN} is the AlGaN permittivity. For simplicity, the AlGaN doping-induced effect on V_{th} can be ignored in the unintentionally doped AlGaN/GaN structure, due to its low intrinsic carrier concentration in the wide-bandgap semiconductor. Therefore, only the effects of the front four terms in Equation (1) were discussed here.

2.1.1. Schottky Barrier Height

The Schottky barrier height under zero bias can be determined by the expression [27]:

$$\varphi_b(T) = \eta\varphi_{EB}(T) - (\eta - 1)[\Delta E_C - E_{F0}(T)] \tag{2}$$

where φ_{EB} and η are the effective Schottky barrier height and ideality factor related to the applied electric field, respectively. They can be deduced from the experimental data using the expression $\ln(I_{GS}/aA^*T^2) = q(V_{GS} - \eta\varphi_{EB})/\eta kT$. Here, I_{GS} and V_{GS} are the gate-to-source current and voltage, a is the gate contact area, and A^* ($= 28.4$ A·cm^{-2}·K^{-2}) is the effective Richardson constant.

2.1.2. Conduction Band Offset

Conduction band offset (ΔE_C) plays an important role in determining the 2DEG density at the AlGaN/GaN interface. ΔE_C at different temperatures can be expressed as below [28,29].

$$\Delta E_C(T) = 0.7\left[E_g^{AlGaN}(T) - E_g^{GaN}(T)\right] \tag{3}$$

where the material bandgap with the different Al composition x can be obtained by

$$E_g^{AlGaN}(T, x) = xE_g^{AlN}(T) + (1 - x)E_g^{GaN}(T) - x(1 - x) \tag{4}$$

$$E_g^{AlN}(T) = 6.31 - 1.80 \times 10^{-3} \times \frac{T^2}{T + 1462} \tag{5}$$

$$E_g^{GaN}(T) = 3.51 - 9.09 \times 10^{-4} \times \frac{T^2}{T + 830} \tag{6}$$

2.1.3. Fermi Energy

The Fermi energy level in GaN semiconductor can be defined by the empirical equation, as below [26].

$$E_{F0}(T) = k_1(T) + k_2(T)n_s^{1/2}(T) + k_3(T)n_s \tag{7}$$

where k_1, k_2, and k_3 are the temperature-dependent parameters and n_s is the sheet charge density.

2.1.4. Polarization Charge

High-density 2DEG will be formed at the AlGaN/GaN interface due to the existence of the large spontaneous and piezoelectric polarization induced electric field that can be determined by the elastic and piezoelectric constants of the materials in the c-axis direction. The sheet charge density induced by the net polarization will be deduced directly by the polarization expression from the published material parameters [28]. Although the net polarization-induced charge density is obviously changed depending on the Al composition and AlGaN barrier thickness, the pyroelectric coefficients that describe the changes in polarization depending on the temperature were found to be very small [30,31]. Chang et al. proved that the effect of pyroelectric coefficients on the channel current at high temperatures is negligible [3]. Therefore, the net polarization charge density was considered to be constant at a temperature less than 150 °C in this work.

2.2. Metal-Insulator-Semiconductor (MIS)-Gate AlGaN/GaN HEMTs

The V_{th} shift occurs more often in MIS-gate HEMTs than in Schottky-gate devices. The dominant reason is the existence of high-density traps at the dielectric/AlGaN interface near the gate. Although it is generally accepted that the V_{th} instability is induced by the combined effects of several trap species that are located in the GaN bulk or dielectric oxide or at the dielectric/AlGaN interface, the influence of the dielectric/AlGaN interface traps should be the most significant, given that it has the highest trap density (10^{12}–10^{13} cm^{-2}) and shortest distance to the 2DEG channel.

The band energy level of these traps plays an important role in determining the V_{th} values of the devices when they are operated at different high temperatures. Moreover, the amount of variation of these traps also affects the consistency of the V_{th} during repeated high-temperature operation. The trap amount will finally decrease and become stable after high-temperature measurements for several rounds. The repeated carrier injection and hopping among the traps due to thermal diffusion or the tunnelling process with the help of an electric field, are considered to be the main physical mechanisms that finally stabilize the trap energy level and amount after repeated high-temperature current–voltage (*I-V*) measurements. Therefore, the effects of the trap energy level and amount at the dielectric/AlGaN interface are included in this section to redefine the V_{th} in the MIS-gate AlGaN/GaN HEMTs. The V_{th} can be derived from the expressions below.

$$V_{th2}(T) = \frac{\varphi_b(T)}{q} - \frac{\Delta E_{C1}(T)}{q} + \frac{E_{F0}(T)}{q} - \frac{\Delta E_{C2}(T)}{q} - \frac{qd\sigma_{pol}(T)}{\varepsilon_{AlGaN}} - \frac{qd_{OX}\sigma_{pol}(T)}{\varepsilon_{OX}} - \frac{qd_{OX}N_{IT}(T)}{\varepsilon_{OX}} \tag{8}$$

$$N_{IT}(T,t) = N_{IT0}(T)\left[1 - k_t \exp\left(-\frac{1}{t}\right)\right] \tag{9}$$

$$N_{IT0}(T) = N_{IT0}(T = 25\ °C) \exp\left[-\frac{\Delta E_{IT}(T)}{kT}\right] \tag{10}$$

$$\tau_n = \frac{1}{NCv\sigma_n} \exp\left(\frac{E_C - E_D}{kT}\right) \tag{11}$$

Here d_{OX} and ε_{OX} are the thickness and dielectric constant of the gate oxide, respectively. ΔE_{C1} and ΔE_{C2} are the conduction band offsets at the AlGaN/GaN and dielectric/AlGaN interfaces, respectively. $N_{IT}(T)$ is the charged trap density at the dielectric/AlGaN interface which is varied with the measurement temperature and operation count at high temperature up to 150 °C. $N_{IT0}(T)$ is the initial interface trap density. k_t is the coefficient to describe the effect of the high-temperature measurement count on the trap amount and t is the repeated measurement rounds. $\Delta E_{IT}(T)$ is the variation value of the trap energy level at different measurement temperatures. N_C, v, σ_n, and E_D are the effective density of states in the conduction band, the thermal velocity of electrons, the capture cross section and the energetic location of the traps, respectively. $E_C - E_D$ = 0.37 eV and $v = 6 \times 10^6$ cm·s^{-1} at AlGaN surface were used in the work [12].

The front six terms in Equation (8) describe the combined effects that originated from the basic physical parameters, i.e., the Schottky barrier, energy band, and polarization-induced charges on the V_{th} variation in the MIS-gate HEMTs, which are similar and consistent with the Schottky-gate case discussed earlier. The term $qd_{OX}\sigma_{pol}(T)/\varepsilon_{OX}$ in Equation (8) indicates an additional effect of the polarization-induced charges on the V_{th} by adding the gate oxide dielectric in the HEMTs. As mentioned, the effect was negligible since the variation of the polarization charge density with the temperature was small. The last term in Equation (8) related to the interface traps will play a key role in how the V_{th} changes with the measurement temperature and repeated rounds. Equation (9) is proposed to govern the amount variation of the charged interface traps, with increasing high temperature measurement rounds. Initially ($t\sim0$), N_{IT} is equal to $N_{IT0}(T)$, corresponding to the case without any high-temperature operation. Then N_{IT} decreased with the increasing measurement count t and finally the trap amount will be constant, and the V_{th} value will be stable from Equation (9). The coefficient k_t is used to distinguish the effects of different measurement temperatures on the trap amount during the repeated high-temperature measurements. In Equation (10), the trap energy level is dependent on the temperature. The effective trap amount can be determined by the variation of the trap energy level with the increasing measurement temperature. In this work, the Shockley-Read-Hall (SRH) model was employed to describe the physical charge-trapping behaviour through the deep defect level in the gap. Several mathematical models were used to determine the temperature-dependent parameters, such as the material bandgap, Fermi level, polarization charge, carrier mobility and carrier lifetimes (or time constant) in the simulation work.

3. Experiments and Results

3.1. Fabrication and Measurement Process

The schematic and optical microscopy image of the typical MIS-gate HEMTs are shown in Figure 1a,b, respectively. The specifics of the device structure are contained in Figure 1a. The fabrication work of the devices began with the mesa isolation by selectively etching the epitaxial layers with 300 nm depth. Ideal source and drain Ohmic contacts were achieved by depositing Ti/Al/Ni/Au (25 nm/125 nm/ 45 nm/55 nm) metal alloys using the E-beam system and annealing at 850 °C for 30 s in N_2. SiO_2 film with 150 nm thickness was deposited using a plasma-enhanced chemical vapour deposition (PECVD) system for surface passivation. Then the gate window with 2 μm length and 200 μm width was defined by photolithography, followed by the removal of the SiO_2 film by wet etching. The Al_2O_3 gate dielectric layer with 15 nm thickness was deposited by the atomic layer deposition (ALD) system. Ni/Au metals were then deposited using the E-beam system for the gate electrode.

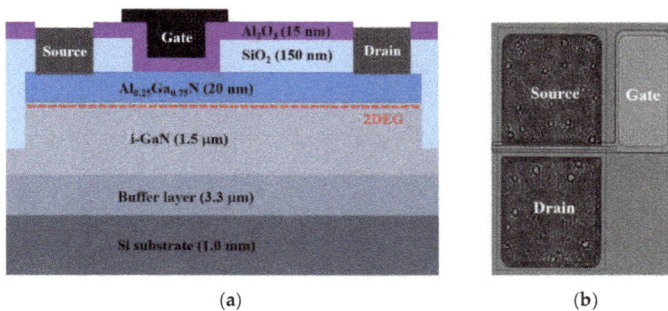

Figure 1. (a) Cross-sectional schematic and (b) optical microscopy image of the fabricated AlGaN/GaN metal–insulator–semiconductor (MIS)-gate high electron mobility transistors (HEMTs).

The *I-V* characteristics of the fabricated devices were measured at gradually increasing temperatures from 25 °C to 150 °C with a step of 25 °C using an Agilent B1505A semi-conductor device analyser (Agilent, Santa Clara, CA, USA). The measurement process was repeated for several rounds, and the testing platform was always cooled down naturally to room temperature before starting a new measurement round. Each measurement point was maintained at a setting temperature for 10 min in ambient air and then the drain current-gate voltage (I_d-V_g) transfer curves were measured. The average measurement time for each *I-V* curve was around 2 s. The device-related physical parameters employed in the modelling and TCAD simulation processes were calibrated by benchmarking the device *I-V* characteristics with the measurement data. The typical parameter values are listed in Table 1 [12,32]. Verification was made by the simulation and laboratory measurement data to support the validity of the proposed model in the paper.

Table 1. Summary of the typical parameters adopted in this work. 2DEG–two dimensional electron gas.

Parameters	Descriptions	Values
n_s	2DEG sheet density	6.7×10^{12} cm^{-2}
σ_{pol}	Sheet density of polarization charges at the AlGaN/GaN interface	1.0×10^{13} cm^{-2}
N_{IT}	Donor-like trap density at the gate dielectric/AlGaN interface	6.5×10^{12} cm^{-2}
μ_n	Electron mobility	1050 cm$^2 \cdot$V$\cdot$s^{-1}
φ_{Ni}	Ni work function	5.1 eV
ε_{OX}	Al$_2$O$_3$ dielectric constant	9.0
ε_{AlGaN}	AlGaN dielectric constant	10.3 (when x = 0.25)
φ_b	Barrier height	1.5 eV for Schottky and 3.7 eV for MIS gates
E_{F0}	Fermi level from GaN conduction band edge	0.39 eV
ΔE_{C2}	Conduction band offset at the Al$_2$O$_3$/AlGaN interface	1.10 eV
ΔE_{C1}	Conduction band offset at the AlGaN/GaN interface	0.36 eV
k_f	Stability coefficient	0.75 at 25 °C; 0.32 at 100 °C; 0.15 at 150 °C
τ_n	Time constant at 25 °C	8.6×10^{-7} s
N_C	Conduction band state density	2.2×10^{18} cm^{-3} for GaN and 4.1×10^{18} cm^{-3} for AlN
σ_n	Capture cross section of the traps	1.0×10^{-13} cm^2 for both GaN and AlN
α	Temperature coefficient	9.09×10^{-4} eV$\cdot$K^{-1} for GaN and 1.80×10^{-3} eV$\cdot$K^{-1} for AlN

3.2. Model Verification and Discussion

3.2.1. Results for Schottky-Gate AlGaN/GaN HEMTs

Figure 2 shows the detailed effects of the main physical parameters that varied with the temperature on the V_{th} stability. The V_{th} shifted slightly towards the negative direction as the temperature increased. The general effect on the V_{th} variation in the Schottky-gate HEMTs is displayed in Figure 2. The variation value of V_{th} was found to be around 0.15 V in the Schottky-gate devices when the temperature changed from 25 °C to 150 °C, which demonstrates that the V_{th} shift in Schottky-gate devices is relatively small.

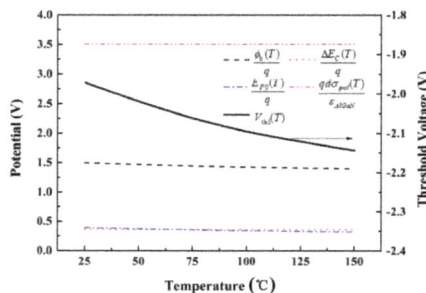

Figure 2. Specific contributions of the main physical parameters to the threshold voltage (V_{th}) variations in the Schottky-gate AlGaN/GaN HEMTs using the analytical model (dash lines). The solid line shows the general V_{th} value that varied with the temperature.

3.2.2. Results for the MIS-Gate AlGaN/GaN HEMTs

Figure 3 shows the measurement of the temperature-dependent DC output characteristics of the fabricated MIS-gate HEMTs. The drain current density decreased with the increasing temperature. This was mainly attributed to the degeneration of the electron mobility in the 2DEG channel induced by the thermal lattice vibration scattering. The source-to-drain channel conductance g_{ds} was found to be 52.3 mS/mm, 46.6 mS/mm, 39.8 mS/mm, 35.6 mS/mm, 30.4 mS/mm, and 27.5 mS/mm, respectively. Figure 4a shows the typical I_d-V_g curves and gate transconductance g_m characteristics of the fabricated MIS-gate, AlGaN/GaN HEMTs. The g_m peak declined by 34.7% when the measurement temperature was up to 150 °C, which suggests a reduction of the device switching frequency.

The V_{th} values were determined by extrapolating the linear portion of the plot of the drain current density (I_d) to the x axis (V_g). The intercept at the voltage axis was defined as the V_{th} in this paper. $V_d = 1$ V was used in the measurement process for the device transfer characteristics. An illustration of V_{th} definition is provided in Figure 4a. Figure 4b displays a trend of the V_{th} data variation with various temperatures. It was found that the measured V_{th} data shifted from -4.1 V to -2.7 V, towards more positive values with the increasing temperature. The big variance likely mainly originated from the combined effects of the change of interface trap number and the shift of the trap energy level during the process of the temperature increasing. More electrons may be captured by high-density ionized donor-like traps at the Al_2O_3/AlGaN interface beneath the gate at high temperatures. Thus, the number of the interface positive charges decreased, which resulted in the shift of the V_{th}. Equation (10) was employed to clearly describe the relationship between the trap density and the operation temperature. Both the calculation results based on the proposed analytical model and the simulation work agree well with the experimental data, which supports the validity of the physical model.

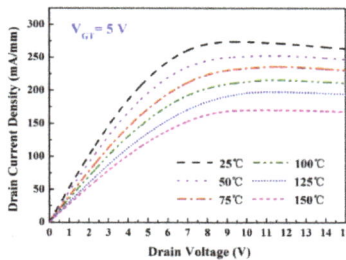

Figure 3. Measurement of the temperature-dependent direct current (DC) output characteristics of the fabricated MIS-gate HEMTs. $V_{GT} = V_g - V_{th} = 5$ V was employed in the measurements.

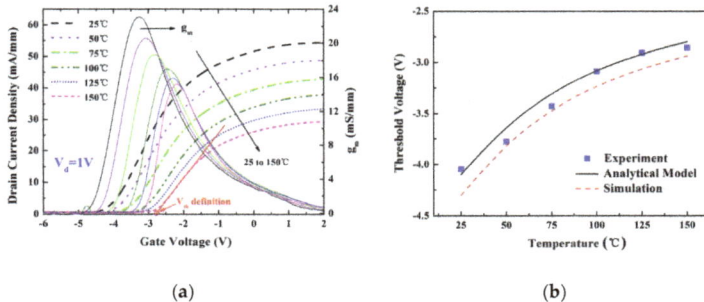

(a)   (b)

Figure 4. (a) Temperature-dependence of the typical device transfer characteristics and gate transconductance g_m in the MIS-gate HEMTs, and (b) the V_{th} data changed with the increasing temperatures in the experimental measurement, analytical calculation using the model, and technology computer-aided design (TCAD) simulation work.

Figure 5 shows the repeated measurement-dependence of the DC output characteristics of the fabricated MIS-gate HEMTs, which demonstrates that the drain current remained almost unchanged even after several rounds of *I-V* measurements. Figure 6 shows the typical I_d-V_g curves and g_m characteristics of the fabricated MIS-HEMTs, dependent on the repeated measurement rounds. The g_m peak changed slightly after several measurement rounds and the variation was less than 7.8%. Figure 7 displays the V_{th} values and the measurement flow. The data were measured at gradually increased temperatures from 25 °C to 150 °C and then the sample was cooled down naturally to room temperature before starting a new measurement round. The measurement process was repeated for several rounds until the V_{th} value was stable. Thus, the repeated measurement dependence of the V_{th} was achieved, as shown in Figure 7. A big variance around 1.1 V was found at the beginning, while the V_{th} value was kept nearly constant and only a small change of 0.1 V was found after several rounds of high-temperature measurements. This indicates that most ionized donor-like traps capture the free electrons and hence the number of charged traps is finally close to constant.

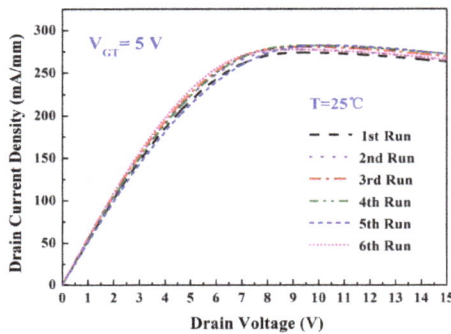

Figure 5. Repeated measurement-dependence of the DC output characteristics in the fabricated MIS-gate HEMTs.

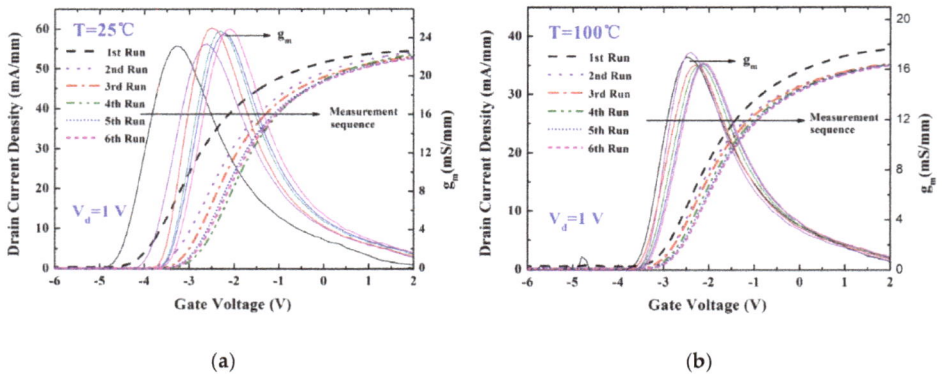

Figure 6. Repeated measurement dependence of the typical device transfer characteristics and gate transconductance g_m in the MIS-gate HEMTs at (**a**) 25 °C and (**b**) 100 °C, respectively.

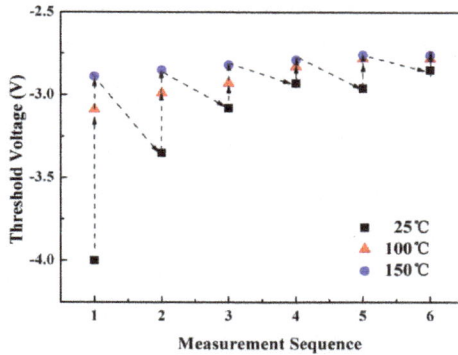

Figure 7. The V_{th} data changed with the repeated measurement rounds in the MIS-gate HEMTs. The data were measured at gradually increasing temperatures from 25 °C to 150 °C and the measurement process was repeated for several rounds. The dotted lines and arrows show the measurement sequence.

Figure 8 shows the employed trap density data at the Al_2O_3/AlGaN interface in the analytical model and TCAD simulation, which are dependent on the measurement temperature and sequence. The amount of variation of these charged traps greatly affected the consistency of the V_{th} during the repeated high-temperature operations. The charged trap amount was assumed to decrease with the increasing measurement rounds and finally became stable in the proposed physical model. The repeated carrier injection and hopping among the traps by the thermal diffusion and/or tunnelling process in the measurements, were considered to be the main physical mechanism responsible for the amount of variation of the charged traps. This process might repair some defects to some extent, and finally stabilizes the charged trap number after repeated high-temperature *I-V* measurements. Future work should explore whether the applied low electric field at high temperatures helps to repair the interface defects around the gate. Equation (9) in the manuscript was employed to govern the amount of variation of the traps with the increasing high-temperature measurement rounds. Figure 9 displays the V_{th} values that changed with the measurement sequence based on the experimental measurement, analytical model, and TCAD simulation work. The results were basically consistent and the average mismatch for the V_{th} value was kept within 5%, which again supports the validity of the physical model.

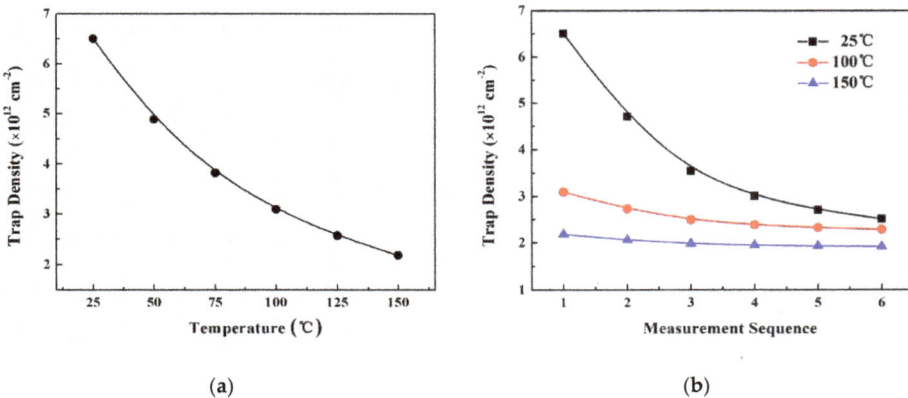

(a)

(b)

Figure 8. Trap density data at the Al_2O_3/AlGaN interface employed in the analytical model and TCAD simulation, dependent on (**a**) the operation temperature and (**b**) measurement sequence.

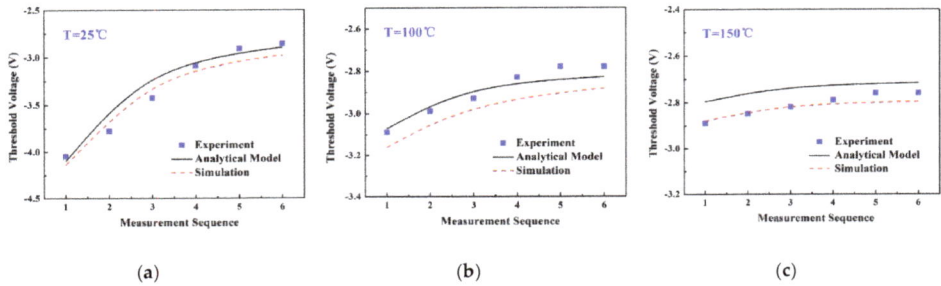

Figure 9. The V_{th} data changed with the measurement sequence from the experimental measurement, analytical calculation using the model, and TCAD simulation work when operated at (**a**) 25 °C, (**b**) 100 °C, and (**c**) 150 °C, respectively.

4. Conclusions

GaN-based HEMT devices were fabricated and measured to investigate the effects of the operation temperature and repeated rounds on the V_{th} stability. The V_{th} analytical model was proposed and developed to study the mechanism of the V_{th} variations in the repeated high-temperature operations. The combined effects of the amount of change of the interface charged traps and the shift of the trap energy level were considered to be the main reason for the V_{th} shift. The validity of the proposed analytical model was verified by experimental measurement and TCAD simulation results. The work can assist the engineers find a better understanding of the V_{th} stability of power devices in practical high-temperature applications.

Author Contributions: Data curation, F.L.; Investigation, Z.S.; Software, Y.C.; Writing—review and editing, H.H.

Funding: This research was funded by the National Science Foundation of China (grant number 51607022), the Fundamental Research Funds for the Central Universities (grant number DUT17LK13), the Major University Science Research Project of Anhui Province (grant number KJ2016SD42), the Open Project Program of Key Lab of Semiconductor Materials Science (grant number KLSMS-1610) and the Open Project Program of Key Laboratory of Nanodevices and Applications (grant number 18JG02) from Chinese Academy of Sciences.

Acknowledgments: The authors are also grateful to the National University of Singapore and Yung C. Liang for his time and help.

Conflicts of Interest: The authors declare no conflict of interest.

References

1. Chen, C.; Liu, X.Z.; Tian, B.L.; Shu, P.; Chen, Y.F.; Zhang, W.L.; Jiang, H.C.; Li, Y.R. Fabrication of Enhancement-Mode AlGaN/GaN MISHEMTs by Using Fluorinated Al_2O_3 as Gate Dielectrics. *IEEE Electron Device Lett.* **2011**, *32*, 1373–1375. [CrossRef]
2. Zhang, Y.H.; Sun, M.; Joglekar, S.J.; Fujishima, T.; Palacios, T. Threshold voltage control by gate oxide thickness in fluorinated GaN metal-oxide-semiconductor high-electron-mobility transistors. *Appl. Phys. Lett.* **2013**, *103*, 033524. [CrossRef]
3. Chang, Y.; Tong, K.Y.; Surya, C. Numerical simulation of current–voltage characteristics of AlGaN/GaN HEMTs at high temperatures. *Semicond. Sci. Technol.* **2005**, *20*, 188. [CrossRef]
4. Campbell, C.F.; Balistreri, A.; Kao, M.Y.; Dumka, D.C.; Hitt, J. GaN Takes the Lead. *IEEE Microw. Mag.* **2012**, *13*, 44–53. [CrossRef]
5. Crupi, G.; Vadala, V.; Colantonio, P.; Cipriani, E.; Caddemi, A.; Vannini, G.; Schreurs, D.M.M.P. Empowering GaN HEMT models: The gateway for power amplifier design. *Int. J. Numer. Model. Electron. Netw. Devices Fields* **2017**, *30*, e2125. [CrossRef]
6. Quaglia, R.; Camarchia, V.; Pirola, M.; Ghione, G. GaN Monolithic Power Amplifiers for Microwave Backhaul Applications. *Electronics* **2016**, *5*, 25. [CrossRef]

7. Nalli, A.; Raffo, A.; Crupi, G.; D'Angelo, S.; Resca, D.; Scappaviva, F.; Salvo, G.; Caddemi, A.; Vannini, G. GaN HEMT Noise Model Based on Electromagnetic Simulations. *IEEE Trans. Microw. Theory* **2015**, *63*, 2498–2508. [CrossRef]

8. Gurpinar, E.; Castellazzi, A. Tradeoff Study of Heat Sink and Output Filter Volume in a GaN HEMT Based Single-Phase Inverter. *IEEE Trans. Power Electr.* **2018**, *33*, 5226–5239. [CrossRef]

9. Li, J.; Mao, S.; Xu, Y.; Zhao, X.; Wang, W.; Guo, F.; Zhang, Q.; Wu, Y.; Zhang, B.; Chen, T.; et al. An Improved Large Signal Model for 0.1 μm AlGaN/GaN High Electron Mobility Transistors (HEMTs) Process and Its Applications in Practical Monolithic Microwave Integrated Circuit (MMIC) Design in W band. *Micromachines* **2018**, *9*, 396. [CrossRef] [PubMed]

10. Bae, K.T.; Lee, I.J.; Kang, B.; Sim, S.; Jeon, L.; Kim, D.W. X-Band GaN Power Amplifier MMIC with a Third Harmonic-Tuned Circuit. *Electronics* **2017**, *6*, 103. [CrossRef]

11. Meneghini, M.; Rossetto, I.; Rizzato, V.; Stoffels, S.; Van Hove, M.; Posthuma, N.; Wu, T.L.; Marcon, D.; Decoutere, S.; Meneghesso, G.; et al. Gate Stability of GaN-Based HEMTs with P-Type Gate. *Electronics* **2016**, *5*, 14. [CrossRef]

12. Huang, H.L.; Sun, Z.H.; Cao, Y.Q.; Li, F.Y.; Zhang, F.; Wen, Z.X.; Zhang, Z.F.; Liang, Y.C.; Hu, L.Z. Investigation of surface traps-induced current collapse phenomenon in AlGaN/GaN high electron mobility transistors with schottky gate structures. *J. Phys. D Appl. Phys.* **2018**, *51*, 345102. [CrossRef]

13. He, J.; Hua, M.; Zhang, Z.; Chen, K.J. Performance and V_{th} Stability in E-Mode GaN Fully Recessed MIS-FETs and Partially Recessed MIS-HEMTs with LPCVD-SiN$_x$/PECVD-SiN$_x$ Gate Dielectric Stack. *IEEE Trans. Electron Devices* **2018**, *65*, 3185–3191. [CrossRef]

14. Jogai, B. Influence of surface states on the two-dimensional electron gas in AlGaN/GaN heterojunction field-effect transistors. *J. Appl. Phys.* **2003**, *93*, 1631–1635. [CrossRef]

15. Horio, K.; Yonemoto, K.; Takayanagi, H.; Nakano, H. Physics-based simulation of buffer-trapping effects on slow current transients and current collapse in GaN field effect transistors. *J. Appl. Phys.* **2005**, *98*, 124502. [CrossRef]

16. Lin, Y.S.; Wu, J.Y.; Chan, C.Y.; Hsu, S.S.H.; Huang, C.F.; Lee, T.C. Square-Gate AlGaN/GaN HEMTs With Improved Trap-Related Characteristics. *IEEE Trans. Electron Devices* **2009**, *56*, 3207–3211. [CrossRef]

17. Brannick, A.; Zakhleniuk, N.A.; Ridley, B.K.; Shealy, J.R.; Schaff, W.J.; Eastman, L.F. Influence of Field Plate on the Transient Operation of the AlGaN/GaN HEMT. *IEEE Electron Device Lett.* **2009**, *30*, 436–438. [CrossRef]

18. Wang, C.S.; Xu, Y.H.; Yu, X.M.; Ren, C.J.; Wang, Z.S.; Lu, H.Y.; Chen, T.S.; Zhang, B.; Xu, R.M. An Electrothermal Model for Empirical Large-Signal Modeling of AlGaN/GaN HEMTs Including Self-Heating and Ambient Temperature Effects. *IEEE Trans. Microw. Theory* **2014**, *62*, 2878–2887. [CrossRef]

19. Crupi, G.; Raffo, A.; Avolio, G.; Schreurs, D.M.M.P.; Vannini, G.; Caddemi, A. Temperature Influence on GaN HEMT Equivalent Circuit. *IEEE Microw. Wirel. Compon. Lett.* **2016**, *26*, 813–815. [CrossRef]

20. Wang, R.N.; Cai, Y.; Chen, K.J. Temperature dependence and thermal stability of planar-integrated enhancement/depletion-mode AlGan/GaN HEMTs and digital circuits. *Solid-State Electron.* **2009**, *53*, 1–6. [CrossRef]

21. Tallarico, A.N.; Stoffels, S.; Posthuma, N.; Magnone, P.; Marcon, D.; Decoutere, S.; Sangiorgi, E.; Fiegna, C. PBTI in GaN-HEMTs with p-Type Gate: Role of the Aluminum Content on V_{th} and Underlying Degradation Mechanisms. *IEEE Trans. Electron Devices* **2018**, *65*, 38–44. [CrossRef]

22. Khan, M.N.; Ahmed, U.F.; Ahmed, M.M.; Rehman, S. An improved model to assess temperature-dependent DC characteristics of submicron GaN HEMTs. *J. Comput. Electron.* **2018**, *17*, 653–662. [CrossRef]

23. Van Hove, M.; Kang, X.W.; Stoffels, S.; Wellekens, D.; Ronchi, N.; Venegas, R.; Geens, K.; Decoutere, S. Fabrication and Performance of Au-Free AlGaN/GaN-on-Silicon Power Devices with Al$_2$O$_3$ and Si$_3$N$_4$/Al$_2$O$_3$ Gate Dielectrics. *IEEE Trans. Electron Devices* **2013**, *60*, 3071–3078. [CrossRef]

24. Johnson, D.W.; Lee, R.T.P.; Hill, R.J.W.; Wong, M.H.; Bersuker, G.; Piner, E.L.; Kirsch, P.D.; Harris, H.R. Threshold Voltage Shift Due to Charge Trapping in Dielectric-Gated AlGaN/GaN High Electron Mobility Transistors Examined in Au-Free Technology. *IEEE Trans. Electron Devices* **2013**, *60*, 3197–3203. [CrossRef]

25. Park, B.R.; Lee, J.G.; Choi, W.; Kim, H.; Seo, K.S.; Cha, H.Y. High-Quality ICPCVD SiO$_2$ for Normally off AlGaN/GaN-on-Si Recessed MOSHFETs. *IEEE Electron Device Lett.* **2013**, *34*, 354–356. [CrossRef]

26. Huque, M.A.; Eliza, S.A.; Rahman, T.; Huq, H.F.; Islam, S.K. Temperature dependent analytical model for current–voltage characteristics of AlGaN/GaN power HEMT. *Solid-State Electron.* **2009**, *53*, 341–348. [CrossRef]

27. Wang, Y.H.; Liang, Y.C.; Samudra, G.S.; Chang, T.F.; Huang, C.F.; Yuan, L.; Lo, G.Q. Modelling temperature dependence on AlGaN/GaN power HEMT device characteristics. *Semicond. Sci. Technol.* **2013**, *28*, 125010. [CrossRef]

28. Ambacher, O.; Foutz, B.; Smart, J.; Shealy, J.R.; Weimann, N.G.; Chu, K.; Murphy, M.; Sierakowski, A.J.; Schaff, W.J.; Eastman, L.F.; et al. Two dimensional electron gases induced by spontaneous and piezoelectric polarization in undoped and doped AlGaN/GaN heterostructures. *J. Appl. Phys.* **2000**, *87*, 334–344. [CrossRef]

29. Vurgaftman, I.; Meyer, J.R. Band parameters for nitrogen-containing semiconductors. *J. Appl. Phys.* **2003**, *94*, 3675–3696. [CrossRef]

30. Ambacher, O.; Majewski, J.; Miskys, C.; Link, A.; Hermann, M.; Eickhoff, M.; Stutzmann, M.; Bernardini, F.; Fiorentini, V.; Tilak, V.; et al. Pyroelectric properties of Al(In)GaN/GaN hetero- and quantum well structures. *J. Phys. Condens. Mater.* **2002**, *14*, 3399–3434. [CrossRef]

31. Shur, M.S.; Bykhovski, A.D.; Gaska, R. Pyroelectric and piezoelectric properties of GaN-based materials. *MRS Proc.* **1999**, *537*. [CrossRef]

32. Huang, H.L.; Liang, Y.C. Formation of combined partially recessed and multiple fluorinated-dielectric layers gate structures for high threshold voltage GaN-based HEMT power devices. *Solid-State Electron.* **2015**, *114*, 148–154. [CrossRef]

micromachines

MDPI

Article

A Breakdown Enhanced AlGaN/GaN Schottky Barrier Diode with the T-Anode Position Deep into the Bottom Buffer Layer

Youlei Sun, Ying Wang *, Jianxiang Tang, Wenju Wang, Yifei Huang and Xiaofei Kuang

The Key Laboratory of RF Circuits and Systems, Ministry of Education, Hangzhou Dianzi University, Hangzhou 310018, China; 161040013@hdu.edu.cn (Y.S.); hztangjx@hdu.edu.cn (J.T.); 162040172@hdu.edu.cn (W.W.); 162040160@hdu.edu.cn (Y.H.); kuangxiaofei@hdu.edu.cn (X.K.)
* Correspondence: wangying01@hdu.edu.cn or wangying7711@yahoo.com

Received: 15 December 2018; Accepted: 22 January 2019; Published: 26 January 2019

check for
updates

Abstract: In this paper, an AlGaN/GaN Schottky barrier diode (SBD) with the T-anode located deep into the bottom buffer layer in combination with field plates (TAI-BBF FPs SBD) is proposed. The electrical characteristics of the proposed structure and the conventional AlGaN/GaN SBD with gated edge termination (GET SBD) were simulated and compared using a Technology Computer Aided Design (TCAD) tool. The results proved that the breakdown voltage (V_{BK}) in the proposed structure was tremendously improved when compared to the GET SBD. This enhancement is attributed to the suppression of the anode tunneling current by the T-anode and the redistribution of the electric field in the anode–cathode region induced by the field plates (FPs). Moreover, the T-anode had a negligible effect on the two-dimensional electron gas (2DEG) in the channel layer, so there is no deterioration in the forward characteristics. After being optimized, the proposed structure exhibited a low turn-on voltage (V_T) of 0.53 V and a specific on-resistance ($R_{ON,sp}$) of 0.32 mΩ·cm^2, which was similar to the GET SBD. Meanwhile, the TAI-BBF FP SBD with an anode-cathode spacing of 5 μm achieved a V_{BK} of 1252 V, which was enhanced almost six times compared to the GET SBD with a V_{BK} of 213 V.

Keywords: T-anode; GaN; buffer layer; anode field plate (AFP); cathode field plate (CFP)

---

1. Introduction

At present, most power semiconductor devices are fabricated from Si materials, but as the process progresses, the performance of Si devices is approaching the material limit. Therefore, wide-bandgap semiconductor materials such as diamond [1,2], SiC [3,4], and GaN [5] have become promising candidates to make high power semiconductor devices. These wide-bandgap semiconductor materials have a high breakdown field [2], high thermal conductivity [6,7], and an extremely low intrinsic carrier concentration at room temperature, which can make power devices with high potential figures of merit [2]. However, GaN power devices such as AlGaN/GaN HEMT and Schottky barrier diodes (SBD) have attracted more attention, as the GaN heterojunction can make them have a faster switching speed, high breakdown voltage, and low on-resistance [8–13]. In this work, an AlGaN/GaN SBD will be investigated in detail. Despite the advantage mentioned above, there are still many unsolved problems before AlGaN/GaN SBD can be used on a large scale, such as the high turn-on voltage, high anode leakage, and low breakdown voltage. By now, numerous approaches like the etching barrier layer [13] and selective Si diffusion [14] have been demonstrated to effectively reduce the onset voltage, but the leakage current is relatively large. In addition, combinations of high/low Schottky barrier metals [15], carbon-doped GaN buffer [16,17], Fe-doped GaN buffer [18], and gated edge termination

(GET) [19,20] have been proposed to suppress the reverse leakage, but the double Schottky barrier metal and GET are still a challenging manufacturing issue, and the doped GaN buffer will affect the device's forward characteristics, especially the on-resistance. The utilization of field plates (FPs), such as anode FPs (AFP) [21] and cathode FPs (CFP), is a simple and effective method to modulate the electric field. In our previous work, vertical FPs (VFP) [22] were proposed for the redistribution of the electric field in the anode–cathode region. However, the effect of a vertical FP is much lower than a lateral FP, and the VFP does not contribute to suppressing the anode leakage current.

In this work, a T-anode located deep into the bottom buffer layer of the SBD in combination with field plates (TAI-BBF FPs SBD) is proposed where the T-anode not only plays the role of VFP, but also introduces the leakage electrons into the bottom buffer layer with high concentration acceptor traps, resulting in the significant suppression of the anode leakage current. As a result, the 2DEG in the channel can be depleted and the space charge region will be extended, which leads to the enhancement of the breakdown voltage. Compared to the previous work, double GaN buffers were selected for the T-anode located deep into the bottom buffer layer of the SBD (TAI-BBF SBD). The first buffer layer with low concentration acceptor traps is called the middle buffer layer, and the second buffer layer with high concentration acceptor traps is called the bottom buffer layer, and can not only suppress the leakage current, but also has a negligible effect on the device's forward characteristics. Moreover, the combination of the T-anode and the bottom buffer layer can cause most of the leakage electrons to be trapped in the bottom buffer. Meanwhile, the electrons in the bottom buffer layer can effectively modulate the device's surface electric field. In addition, the GET is integrated into the Schottky contact (SC) serving as the anode [19,20] to reduce the turn-on voltage. The AFP along with the CFP are located separately at the anode and cathode, which can modulate the electric field distribution in the anode–cathode region, which in turn results in an appreciable V_{BK} improvement. More importantly, the proposed structure has a minor influence on the forward characteristics, thus obtaining a huge breakdown voltage while maintaining a low $R_{ON,sp}$.

2. Device Structure and Simulation Model

Devices with identical physical dimensions consisting of a conventional GET SBD [20], a SBD with gated edge termination in combination with field plates (GET FPs SBD), TAI-BBF SBD, and a TAI-BBF FPs SBD are shown in Figure 1.

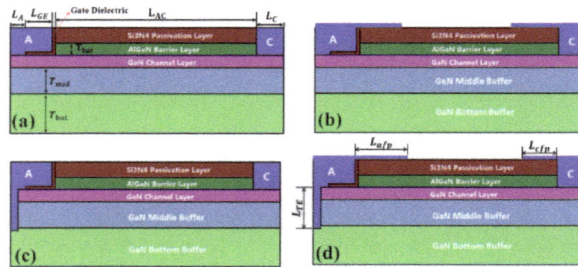

Figure 1. Cross-section of: (**a**) Schottky barrier diode (SBD) with gated edge termination (GET SBD), (**b**) SBD with gated edge termination in combination with field plates (GET FPs SBD), (**c**) T-anode located deep into the bottom buffer layer of the SBD (TAI-BBF SBD), (**d**) T-anode located deep into the bottom buffer layer of the SBD in combination with field plates (TAI-BBF FP SBD).

These architectures in the simulation included a passivation layer, an unintentionally-doped AlGaN barrier layer/GaN channel layer, and a buffer layer. Si_3N_4 of the size 100 nm was selected as the passivation layer [23] and the AlGaN contained a 0.25 Al mole part. In addition, the dielectric layer in the GET was also Si_3N_4 and the thickness was set to 20 nm. The passivation layer and $Al_{0.25}Ga_{0.75}N$ barrier layer under the anode were etched completely in order to achieve a low V_T. The proposed device had an additional T-anode with an initial length of 0.445 µm connecting the Schottky contact with the bottom buffer and two field plates located at the anode and cathode. The buffer layer was divided into a middle buffer and a bottom buffer [17]. The donor impurity would inevitably be introduced in the process of device growth, so the donor concentration in all GaN layers was set to 1×10^{16} cm^{-3} [17] to emulate the donor impurity. In order to reduce the buffer layer leakage, acceptor type traps were doped into both the middle buffer layer and the bottom buffer layer. The effective concentration of the acceptor type traps of 2×10^{16} cm^{-3} [10] and 4×10^{18} cm^{-3} [16,17] were selected to dope into the middle and bottom buffer layers, respectively, in order to neutralize the dopants and reduce the buffer layer leakage. Therefore, the middle and bottom buffer layers became slight p-type and heavy p-type layers, and the density of the hole provided by the two buffer layers could reach 2×10^{16} and 4×10^{18} cm^{-3} [16,24], respectively. The middle buffer close to the channel layer could suppress the leakage current effectively, but it had an imperceptible impact on the 2DEG because of the slight p-type. In contrast, the heavy p-type bottom buffer layer combining with the T-anode could play a leading role in reducing the anode leakage current. Meanwhile, the bottom buffer layer had little effect on the 2DEG, due to its distance from the channel layer. In addition, the energy level of the acceptor trap was set at $E_V + 0.9$ eV [17,25,26] and the cross-section σ_n was selected at 1.3×10^{-14} cm^{-2} [17] in this simulation, because in the previous report [25,26], the energy level of the acceptor trap was defined at $E_V + 0.86$ eV to $E_V + 0.93$ eV. Other device parameters along with their values used in the simulation are listed in Table 1.

The Sentaurus software was selected for 2D numerical simulation. Some necessary physics models [27,28] were adopted, such as mobility models, the Shockley–Read–Hall recombination model, and a polarization model. Furthermore, in the AlGaN/GaN SBD, tunneling leakage at the Schottky junction played the leading role in the reverse leakage current [29], so a nonlocal tunneling model [27] was set at the Schottky contact, and the work function of Schottky anode was defined to be 4.6 eV [20].

Table 1. Major optimized parameters of the proposed structure.

Parameter	Value
Anode length (L_A)	0.2 µm
Gated edge length (L_{GE})	0.45 µm
Cathode length (L_C)	0.1 µm
Anode–cathode spacing (L_{AC})	5 µm
T-anode length (L_T)	0.445–0.945 µm
Barrier layer thickness (T_{bar})	25 nm
Channel layer thickness (T_{chan})	50 nm
Bottom buffer layer thickness (T_{bot})	1.6 µm
Middle buffer layer thickness (T_{mid})	0.4 µm
Cathode field plate length (L_{cfp})	0.6 µm
Anode field plate length (L_{afp})	0.83 µm

Fabrication Process

In order to explain how the proposed device could be implemented, a brief schematic of the fabrication process steps is presented in Figure 2.

The corresponding description of these process steps is summarized as follows.

(a) Metal organic chemical vapor deposition (MOCVD) was adopted to grow the base structure including the GaN bottom buffer layer, the GaN middle buffer layer, the GaN channel layer, and the AlGaN barrier layer, and the Si$_3$N$_4$ passivation layer was then deposited using plasma-enhanced chemical vapor deposition (PECVD) [30].

(b) The Si$_3$N$_4$ passivation layer, AlGaN barrier layer, GaN channel layer, and GaN buffer layer were etched through inductively coupled plasma reactive ion etching (ICP RIE), using a BCl$_3$/Cl$_2$ gas mixture [30].

(c)   The Si$_3$N$_4$ layer was deposited on the anode region using plasma-enhanced chemical vapor deposition (PECVD) [30].

(d)   ICP RIE was adopted to etch the Si$_3$N$_4$ layer of the anode region and kept at a thickness of 20 nm for the Si$_3$N$_4$ layer at the right and bottom side.

(e)   A Ti/Al/Ni/Au ohmic metal was deposited using e-beam evaporation on the cathode, followed by rapid thermal annealing at 800 °C for 30 s in N$_2$ ambient [12]. Lastly, the Schottky metal stack of Ni/Au (40 nm/350 nm) was deposited [12].

The above fabrication process was relatively easy to implement, the only difficulty was that the depth of the groove was not well controlled when etching the T-anode groove. Fortunately, when the T-anode reached 0.445 μm, the breakdown voltage almost reached the saturation value, so there was a certain fault tolerance to the depth of the etched groove.

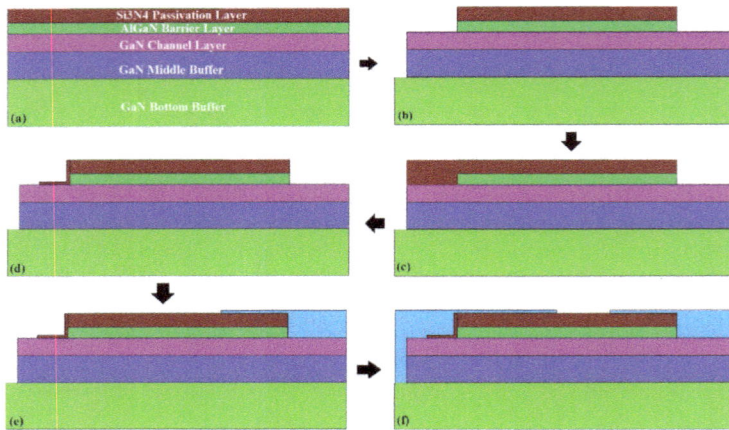

Figure 2. Schematic of the fabrication process steps for the proposed structure.

3. Results and Discussion

3.1. Forward and Reverse Characteristics

The breakdown and forward characteristics of the GET SBD, GET FPs SBD, TAI-BBF SBD, and TAI-BBF FPs SBD are shown in Figure 3.

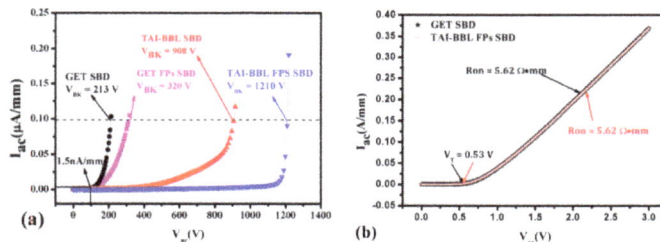

Figure 3. (**a**) The breakdown characteristics of the GET SBD, GET FPs SBD, TAI-BBF SBD and TAI-BBL FPs SBD; (**b**) The forward characteristics of the GET SBD and TAI-BBL FPs SBD.

The breakdown criterion of all the devices was when the anode leakage current reached 0.1 μA/mm, as shown in Figure 3a. In contrast to GET SBD, with a V_{BK} of 213 V, the V_{BK} was enhanced to 320 V in the GET FP SBD, implying that the improvement of V_{BK} was inconspicuous.

However, the V_{BK} was improved to 908 V in the TAI-BBF SBD and further promoted to 1210 V in TAI-BBF FP SBD. In addition, the leakage current reached 1.5 nA/mm when the reverse voltage was 100 V, which was consistent with the experimental results in [20]. The voltage of the forward current reaching 1 mA/mm was defined as the V_T. As shown in Figure 3b, the V_T and R_{on} of the proposed structure were 0.53 V and 5.62 $\Omega\cdot$mm, similar in magnitude to the GET SBD, indicating that the proposed structure showed significant improvement in the reverse characteristics while maintaining the forward characteristics.

3.2. Equipotential Line and Horizontal Electric Field Distribution

The equipotential line and horizontal electric field distribution of four devices when the breakdown occurred are described in Figure 4 to account for the enhancement of the breakdown characteristics. In the GET SBD, the equipotential lines from the anode to point A were very compact. They became sparse after point A, indicating that there were still a large number of electrons in the channel and there was little space charge created from point A to the cathode while the breakdown took place. In the GET FP SBD, the AFP and CFP were added to the GET SBD to modulate the electric field focused on the right of anode and the left of the cathode. However, the V_{BK} of the GET FP SBD only achieved minor improvement. As the electrons in the channel were not being depleted, the FPs did not work out as desired. In contrast to the GET SBD, the introduction of a T-anode in the TAI-BBF SBD resulted in the direct connection of the anode to the bottom buffer layer. A bottom buffer layer with a substantial amount of acceptor traps could help deplete the 2DEG in the channel to expand the space charge region, which could make the equipotential lines more compact throughout the anode–cathode region. Furthermore, the AFP and CFP were placed on the anode and cathode separately in order to make the equipotential line more uniform and denser throughout the whole TAI-BBF FP SBD, as shown in Figure 4d. As a result, the V_{BK} in the TAI-BBL FP SBD was promoted further compared to the GET SBD and TAI-BBL SBD.

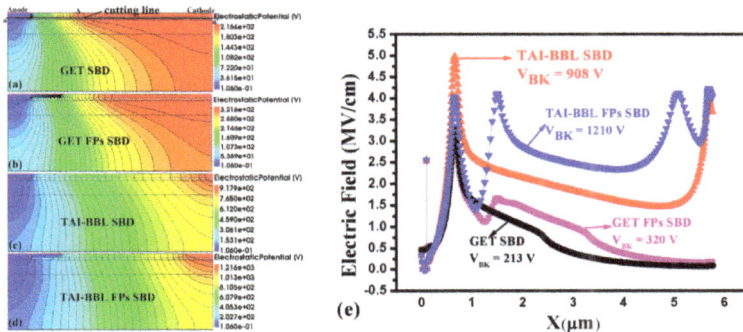

Figure 4. The equipotential line profile of: (**a**) GET SBD, (**b**) GET FP SBD, (**c**) TAI-BBF SBD, (**d**) TAI-BBF FP SBD when breakdown occurs, (**e**) The corresponding horizontal electric field distribution in the 2DEG area.

A horizontal cutting line aa' at 1 nm below the channel of the three devices was carried out to obtain the horizontal electric field distribution, as depicted in Figure 4e. In the GET SBD, the electric field peak occurred at the GET and the field declined sharply to a very small value on the right. In the GET FP SBD, a new field peak occurred to the right of the GET, but the value of the field peak was very small and it then dropped to a small value. In the TAI-BBF SBD, there was another electric field peak at the cathode because almost all of the 2DEG in the channel was depleted by the bottom buffer layer via T-anode, which resulted in the flattening of the electric field contribution in the anode–cathode region, consequently improving V_{BK}. However, the electric field peak at the anode was too high, which resulted in the GET being punctured in advance. In the TAI-BBF FP SBD, the electric field at the GET declined and a new electric field peak appeared below the AFP and CFP due to electric field modulation. After the optimization described above, the V_{BK} became further enhanced as the electric field distributed more uniformly over the anode–cathode area.

3.3. Electron Concentration Distribution

The electron concentration distribution of the GET SBD and TAI-BBL FP SBD when the breakdown occurred are depicted in Figure 5 to explain the role of the T-anode. In the GET SBD, the electron in the channel from the anode to point A was depleted, but from point A to the cathode it was not, as can be seen at the black circle. In addition, there was a very low electron density in the buffer layers, especially the bottom buffer layer, which meant that the high concentration of traps in the buffer layers did not capture the leakage electron as desired. Therefore, there was no space charge generated from point A to the cathode to bear the breakdown voltage, due to the heavy leakage current when the breakdown occurred. In contrast, in the TAI-BBL FP SBD, the electron in the channel was fully depleted, as can be seen in Figure 5c. Moreover, the buffer layers were full of the leakage electron, which implied that the high concentration traps in the buffer layers captured the leakage electron effectively with the assistance of the T-anode. Consequently, the space charge region was extended from point A to the cathode in the TAI-BBL FP SBD resulting in the breakdown voltage being enhanced.

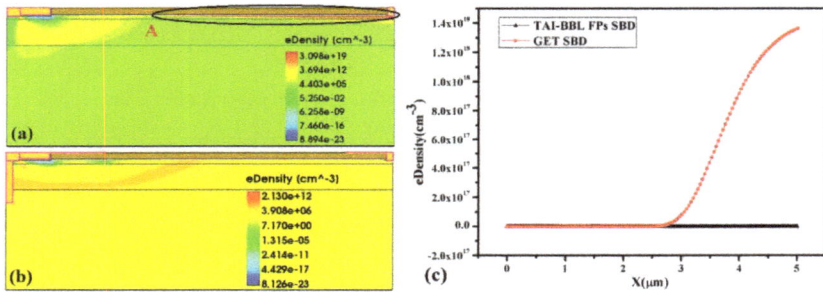

Figure 5. The electron concentration distribution of: (**a**) GET SBD, (**b**) the TAI-BBL FP SBD when the breakdown occurs, (**c**) the corresponding curve distribution.

3.4. The Path of the Leakage Current and the Vertical Electric Field Distribution

The path of the anode leakage current due to the tunneling process [20] and the vertical electric field distribution below the SC during breakdown were generated in order to explain the function of the T-anode directly, as shown in Figure 6.

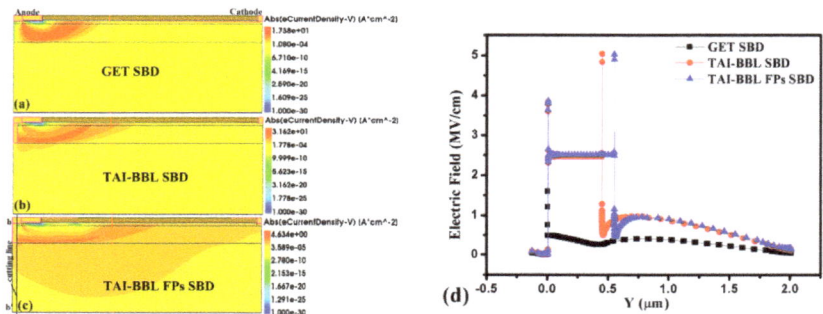

Figure 6. The path of the anode leakage current of: (**a**) GET SBD, (**b**) TAI-BBF SBD, (**c**) TAI-BBF FP SBD, (**d**) the corresponding vertical electric field distribution under the SC.

In the GET SBD, the leakage current tunneled from the SC through the channel and middle buffer layer, and then to the cathode. Thus, the bottom buffer layer had little effect on suppressing the leakage current. In contrast, a part of the leakage current still tunneled through the channel and middle buffer layer to the cathode, but most of the leakage current tunneled from the terminal of the T-anode through the bottom buffer layer to the cathode in the TAI-BBL SBD and the TAI-BBL FP SBD. Therefore, the leakage was significantly suppressed and the V_{BK} was enhanced. A cutting line bb′ of the vertical electric field distribution for all devices was made in the middle of the SC during the breakdown, as shown in Figure 6c. Compare to the GET SBD, the value of the electric field in the vertical orientation was much larger in the TAI-BBF SBD and TAI-BBF FP SBD, which corroborates that the T-anode played the role of a vertical plate [22], so that the electrons were more able to tunnel from the termination of the T-anode into bottom buffer layer. As a result, the anode leakage current was suppressed and the V_{BK} was enhanced. However, if the T-anode only reached the channel layer or the middle buffer layer, the bottom buffer layer would not work as desired and the V_{BK} would decrease, as most of the electrons would tunnel from the T-anode through into the channel and middle buffer layer. Hence, the concentration of electrons in the middle buffer of the TAI-BBL SBD and TAI-BBL FP SBD were much higher than that of the GET SBD, which resulted in the device being punched through in advance.

4. Parameter Optimization

The dependence of the breakdown characteristic on the length of the T-anode and the concentration of the acceptor traps in the bottom buffer layer are plotted in Figure 7. Initially, the length of the T-anode from 0 to 0.3 μm was far from the bottom buffer layer, resulting in a low V_{BK}. When the length of the T-anode exceeded 0.3 μm, the V_{BK} began to increase sharply. Eventually, the V_{BK} achieved a saturation value when the length of T-anode reached 0.845 μm. Thus, the V_{BK} depended on the distance between the T-anode and the bottom buffer layer, as is evident from Figure 6. The concentration of acceptor traps versus the V_{BK} are also shown in Figure 7. However, when the concentration of the acceptor traps of bottom buffer layer increased from 3×10^{18} cm^{-3} to 6×10^{18} cm^{-3}, the V_{BK} remained constant. With the doping concentration of the acceptor traps of 6×10^{18} cm^{-3}, $V_{BK} = 1252$V was obtained for a T-anode length of 0.845 μm. In particular, changing the concentration of the acceptor traps of the bottom buffer layer had a negligible effect on the forward characteristics, as the bottom buffer layer was far from the GaN channel layer.

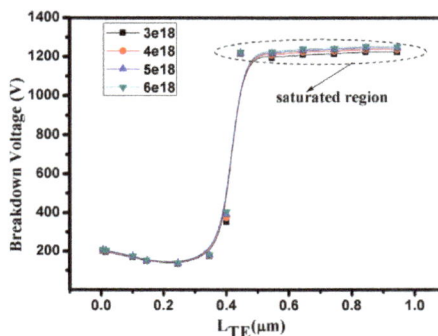

Figure 7. Dependence of the breakdown characteristics on the length of the T anode and the concentration of the acceptor traps of the bottom buffer layer.

5. Conclusions

The function of the T-anode and the bottom buffer was discussed comprehensively in this work. The simulation results showed that the bottom buffer layer with high concentration acceptor traps was able to suppress the anode leakage current effectively via the T-anode, and the T-anode along with the AFP and CFP made the electric field contribution more uniform all over the anode–cathode region. The forward and breakdown characteristics of the GET SBD and the proposed structure were simulated and compared, demonstrating that the proposed structure was able to withstand a larger breakdown voltage while maintaining similar forward characteristics. Finally, a TAI-BBL FP SBD with L_{AC} = 5 μm achieved a V_{BK} of 1252V and a $R_{ON,sp}$ of 0.32 mΩ·cm^2, corresponds to the V_{BK} of 213V and $R_{ON,sp}$ of 0.32 mΩ·cm^2 of the GET SBD. This implies that a satisfactory trade-off between $R_{ON,sp}$ and V_{BK} was obtained in the proposed structure.

Author Contributions: Investigation—Y.S.; Methodology—Y.W.; Software—X.K.; Supervision—W.W., J.T. and Y.H.; Writing original draft—Y.S.

Funding: This work was supported in part by the Excellent Youth Foundation of Zhejiang Province of China (No. p LR17F040001) and in part by the National Natural Science Foundation of China (No. 61774052).

Conflicts of Interest: The authors declare no conflict of interest.

References

1. Holmes, J.; Dutta, M.; Koeck, F.A.; Benipal, M.; Brown, J.; Fox, B.; Hathwar, R.; Johnson, H.; Malakoutian, M.; Saremi, M.; et al. A 4.5-μm PIN diamond diode for detecting slow neutrons. *Nucl. Instrum. Methods Phys. Res. Sect. A* **2018**, *903*, 297–301. [CrossRef]
2. Saremi, M.; Hathwar, R.; Dutta, M.; Koeck, F.A.M. Analysis of the reverse I-V characteristics of diamond-based PIN diodes. *Appl. Phys. Lett.* **2017**, *111*, 043507. [CrossRef]
3. Mahabadi, S.E.J.; Moghadam, H.A. Comprehensive study of a 4H–SiC MES–MOSFET. *Phys. E* **2015**, *74*, 25–29. [CrossRef]
4. Moghadam, H.; Dimitrijev, S.; Han, J.; Han, J.; Haasmann, D.; Aminbeidokhti, A. Transient-current method for measurement of active near-interface oxide traps in 4H-SiC MOS capacitors and MOSFETs. *IEEE Trans. Electron Devices* **2015**, *62*, 2670–2674. [CrossRef]
5. Shen, X.Q.; Matsuhata, H.; Okumura, H. Reduction of the threading dislocation density in GaN films grown on vicinal sapphire (0001) substrates. *Appl. Phys. Lett.* **2005**, *86*, 021912. [CrossRef]
6. Anthony, T.R.; Banholzer, W.F.; Fleischer, J.F.; Wei, L.; Kuo, P.K.; Thomas, R.L.; Pryor, R.W. Thermal diffusivity of isotopically enriched ^{12}C diamond. *Phys. Rev. B* **1990**, *42*, 1104–1111. [CrossRef]
7. Liu, J.; Ohsato, H.; Wang, X.; Liao, M.; Koide, Y. Design and fabrication of high-performance diamond triple-gate field-effect transistors. *Sci. Rep.* **2016**, *6*, 1–8. [CrossRef]
8. Min, W.H.; Seung, C.L.; Young, H.C.; Soo, S.K.; Min, K.H. New GaN Schottky barrier diode employing a trench on AlGaN/GaN heterostructure. *Superlattices Microstruct.* **2006**, *40*, 567–573. [CrossRef]
9. Zhu, M.; Song, B.; Qi, M.; Hu, Z.Y.; Nomoto, K.; Yan, X.D.; Cao, Y.; Johnson, W.; Kohn, E.; Jena, D.; et al. 1.9-kV AlGaN/GaN Lateral Schottky Barrier Diodes on Silicon. *IEEE Electron Device Lett.* **2015**, *36*, 375–377. [CrossRef]
10. Jie, H.; Steve, S.; Ming, Z.; Andrea, N.T.; Isabella, R.; Matteo, M.; Xuan, W.K.; Benoit, B.; Denis, M.; Ben, K.; et al. Time-dependent breakdown mechanisms and reliability improvement in edge terminated AlGaN/GaN Schottky diodes under HTRB tests. *IEEE Electron Device Lett.* **2017**, *38*, 371–374. [CrossRef]
11. Jun, M.; Elison, M. High-voltage and low-leakage AlGaN/GaN tri-anode Schottky diodes with integrated tri-gate transistors. *IEEE Electron Device Lett.* **2017**, *38*, 83–86. [CrossRef]
12. Chuan, W.T.; Kai, P.W.; Yi, W.L.; Shawn, S.H.H. 2.07-kV AlGaN/GaN Schottky barrier diodes on silicon with high Baliga's figure-of-merit. *IEEE Electron Device Lett.* **2016**, *37*, 70–73. [CrossRef]
13. Treidel, E.B.; Oliver, H.; Rimma, Z.; Andreas, W.; Chafik, M.; Joachim, W.; Günther, T. Fast-Switching GaN-Based Lateral Power Schottky Barrier Diodes With Low Onset Voltage and Strong Reverse Blocking. *IEEE Electron Device Lett.* **2012**, *33*, 357–359. [CrossRef]
14. Yi, W.L.; Yu, S.L.; Jui, M.Y.; Chih, H.C.; Shawn, S.H.H. AlGaN/GaN Schottky Barrier Diodes on Silicon Substrates With Selective Si Diffusion for Low Onset Voltage and High Reverse Blocking. *IEEE Electron Device Lett.* **2013**, *34*, 981–983. [CrossRef]
15. Chang, T.F.; Huang, C.F.; Yang, T.Y.; Chiu, C.W.; Huang, T.Y.; Lee, K.Y.; Zhao, F. Low turn-on voltage dual metal AlGaN/GaN Schottky barrier diode. *Solid-State Electron.* **2015**, *105*, 12–15. [CrossRef]

16. Eldad, B.T.; Frank, B.; Oliver, H.; Eunjung, C.; Joachim, W.; Günther, T. AlGaN/GaN/GaN:C Back-Barrier HFETs With Breakdown Voltage of Over 1 kV and Low $R_{ON} \times A$. *IEEE Trans. Electron Devices* **2010**, *57*, 3050–3058. [CrossRef]

17. Xin, L.; Ying, W.; Fei, C.; Cheng, H.Y.; Xin, X.F. A breakdown enhanced AlGaN/GaN MISFET with source-connected P-buried layer. *Superlattices Microstruct.* **2017**, *112*, 517–527. [CrossRef]

18. Chiu, H.C.; Chen, S.C.; Chiu, J.W.; Li, B.H.; Wang, H.Y.; Peng, L.Y.; Wang, H.C.; Hsueh, K.P. AlGaN/GaN Schottky barrier diodes on silicon substrates with various Fe doping concentrations in the buffer layers. *Microelectron. Reliab.* **2018**, *83*, 238–241. [CrossRef]

19. Jing, N.G.; Mao, J.W.; Rui, Y.Y.; Shao, F.L.; Cheng, P.W.; Jin, Y.W.; Wen, G.W.; Yi, L.H.; Yu, F.J.; Bo, S. Schottky-MOS Hybrid Anode AlGaN/GaN Lateral Field-Effect Rectifier With Low Onset Voltage and Improved Breakdown Voltage. *IEEE Trans. Electron Devices* **2017**, *38*, 1425–1428. [CrossRef]

20. Jie, H.; Steve, S.; Silvia, L.; Brice, D.J.; Nicolò, R.; Andrea, N.T.; Dirk, W.; Shu, Z.Y.; Benoit, B.; Guido, G.; et al. Statistical Analysis of the Impact of Anode Recess on the Electrical Characteristics of AlGaN/GaN Schottky Diodes With Gated Edge Termination. *IEEE Trans. Electron Devices* **2016**, *63*, 3451–3458. [CrossRef]

21. Tang, C.; Xie, G.; Zhang, L.; Guo, Q.; Wang, T.; Sheng, K. Electric field modulation technique for high-voltage AlGaN/GaN Schottky barrier diodes. *Chin. Phys. B* **2013**, *22*, 106107:1–106107:6. [CrossRef]

22. Ying, W.; Zhi, Y.L.; Yue, H.; Xin, L.; Jun, P.F.; Ya, C.M.; Cheng, H.Y.; Fei, C. Evaluation by Simulation of AlGaN/GaN Schottky Barrier Diode (SBD) With Anode-Via Vertical Field Plate Structure. *IEEE Trans. Electron Devices* **2018**, *65*, 2552–2557. [CrossRef]

23. Hua, M.; Zhang, Z.; Wei, J.; Lei, J.; Tang, G.; Fu, K.; Cai, Y.; Zhang, B.; Chen, K.J. Integration of LPCVD-SiNx gate dielectric with recessed-gate E-mode GaN MIS-FETs: Toward high performance, high stability and long TDDB lifetime. In Proceedings of the IEEE International Electron Devices Meeting (IEDM), San Francisco, CA, USA, 3–7 December 2016; p. 10. [CrossRef]

24. Kentaro, H.; Hiroshi, O.; Hirohumi, T.; Tohru, N.; Tomoyoshi, M. Junction-Barrier Schottky Diodes Fabricated with Very Thin Highly Mg-Doped p + -GaN(20 nm)/n-GaN Layers Grown on GaN Substrates. In Proceedings of the IEEE International Meeting for Future of Electron Devices, Kansai (IMFEDK), Kyoto, Japan, 29–30 June 2017; pp. 50–51. [CrossRef]

25. Honda, U.; Yamada, Y.; Tokuda, Y.; Shiojima, K. Deep levels in n-GaN doped with carbon studied by deep level and minority carrier transient spectroscopies. *Jpn. J. Appl. Phys.* **2011**, *51*, 827–831. [CrossRef]

26. Lyons, J.L.; Janotti, A.; Walle, C. Carbon impurities and the yellow luminescence in GaN. *Appl. Phys. Lett.* **2010**, *97*, 152108:1–152108:3. [CrossRef]

27. *Sentaurus Device User Guide Version I-2013.12*; Synopsys Inc.: Mountain View, CA, USA, 2013.

28. Stephan, S.; Axel, E.; Tommaso, C.; Denis, M.; Steve, S.; Benoit, B. TCAD methodology for simulation of GaN-HEMT power devices. In Proceedings of the IEEE 26th International Symposium on Power Semiconductor Devices IC's (ISPSD), Waikoloa, HI, USA, 15–19 June 2014; pp. 257–260. [CrossRef]

29. Hu, J.; Lenci, S.; Stoffels, S.; Jaeger, B.D.; Groeseneken, G.; Decoutere, S. Leakage-current reduction and improved on-state performance of Au-free AlGaN/GaN-on-Si Schottky diode by embedding the edge terminations in the anode region. *Phys. Status Solidi C* **2014**, *11*, 862–865. [CrossRef]

30. Hyun, S.L.; Dong, Y.J.; Youngrak, P.; Jeho, N.; Hyun, G.J.; Hyoung, S.L.; Chi, H.J.; Junbo, P.; Sang, O.R.; Sang, C.K.; et al. 0.34 V_T AlGaN/GaN-on-Si Large Schottky Barrier Diode With Recessed Dual Anode Metal. *IEEE Electron Device Lett.* **2015**, *36*, 1132–1134. [CrossRef]

MDPI

St. Alban-Anlage 66

4052 Basel

Switzerland

Tel. +41 61 683 77 34

Fax +41 61 302 89 18

www.mdpi.com

Micromachines Editorial Office

E-mail: micromachines@mdpi.com

www.mdpi.com/journal/micromachines

www.ingramcontent.com/pod-product-compliance
Lightning Source LLC
Chambersburg PA
CBHW051910210326
41597CB00033B/6098